Environmental Ethics, Ecological Theology, and Natural Selection

COLUMBIA SERIES IN SCIENCE AND RELIGION

COLUMBIA SERIES IN SCIENCE AND RELIGION

The Faith of Biology and the Biology of Faith: Order, Meaning, and Free Will in Modern Medical Science
Robert Pollack

Buddhism and Science: Breaking New Ground
B. Alan Wallace, editor

The Columbia Series in Science and Religion consists of peer-reviewed scholarly and general interest titles that probe salient issues relating to science and religion. The series is a forum for the examination of issues that lie at the boundary of these two complementary ways of comprehending the world and our place in it. By examining the intersections between one or more of the sciences and one or more religions, the CSSR hopes to stimulate dialogue and encourage understanding.

Environmental Ethics, Ecological Theology, and Natural Selection

Lisa H. Sideris

Columbia University Press • New York

Columbia University Press
Publishers Since 1893
New York   Chichester, West Sussex

Library of Congress Cataloging-in-Publication Data
Sideris, Lisa H.
Environmental ethics, ecological theology, and natural selection / Lisa H. Sideris.
p. cm.
Includes bibliographical references and index.
ISBN 0-231-12660-3 (acid-free paper) — ISBN 0-231-12661-1 (pbk. : acid-free paper)
1. Environmental ethics. 2. Natural selection. 3. Human ecology—Religious aspects. I. Title.

GE42.S43 2003
179′.1—dc21
2002041713

Casebound editions of Columbia University Press books
are printed on permanent and durable acid-free paper.
Printed in the United States of America
c 10 9 8 7 6 5 4 3 2 1
p 10 9 8 7 6 5 4 3 2 1

CONTENTS

ACKNOWLEDGMENTS

Throughout this work I critique certain uses, and overuses, of terms such as *interdependence* and *interconnectedness*, yet I fully appreciate the interdependent and systemic nature of academic work. I wish to express my gratitude and respect for this particular form of interdependence and the many people who made it possible. This project began as part of my doctoral work in religious studies at Indiana University. Since then it has followed me, or perhaps I have followed it, from town to town and through several transitions. Along the way many individuals provided critical feedback and moral support.

I am grateful to Robert Wuthnow and William Howarth for the opportunity to spend a fellowship year at Princeton University's Center for the Study of Religion, as part of the center's project on Darwin and religion. I profited from conversations with numerous people, both at the center and in the Princeton community as a whole. I particularly wish to thank Charles Mathewes, Marie Griffith, and Max Stackhouse for their comments and encouragement, as well as participants in a workshop at the center where I presented parts of this work.

I am indebted to Holmes Rolston and William French for careful reading and extensive comments on an earlier draft. Professor Rolston also alerted me to recent arguments on suffering, theodicy, and evolution (among other topics), and for that I am very grateful. Since its inception this project has been influenced by Rolston's scholarship, and my appreciation for his work has deepened over the years. I know of no other scholar who has explored the complex and controversial intersection of science, religion, and environmental ethics with such thoroughness and thoughtfulness. Thanks also to my current colleagues at McGill University, particularly Greg Mikkelson, for clarification of certain ecological and ecosystem concepts. Wendy Lochner and Susan Pensak attempted to rein in and lend clarity to my often repetitive and garrulous prose. As always, whatever errors and shortcomings remain in the text—scientific, theological, ethical, or aesthetic—are my responsibility.

I would like to thank my environmental ethics class (2001–2002) at Pace University in Manhattan and Pleasantville for many insightful discussions of environmental issues, bits and pieces of which have surely made their way into this book. Their enthusiasm for the subject helped to refocus my attention on environmental ethics in the dark weeks, and months, following September 11 when it was often difficult to recall why we were there or why the subject still mattered. For better or for worse that experience has left me more attuned than ever to the reality of suffering and the inescapability of human frailty, finitude, and dependence, especially my own. Along these lines, I am grateful for the patience, compassion, and solidarity shown by several good friends and colleagues during a difficult time, especially Carol Burdick, Courtney Campbell, Robert Chapman, Penny Edgell, Richard Miller, Greg Otto, and Donna Rhiver. Last, but certainly not least, I have been fortunate to have the affection and truly good-natured companionship of two nonhuman friends, Lovums and, more recently, Zoe, both of whom are glad to see this book completed so that I can pay more attention to them.

Montreal
December 2002

Environmental Ethics, Ecological Theology, and Natural Selection

INTRODUCTION

Charles Darwin once said that going public with his theory of natural selection was like confessing to murder. Despite the common assumption that we live in a post-Darwinian world, the perennial resurgence of creationism serves as a reminder that Darwinian theory cannot be taken for granted as a shared understanding of the origin of humans and the mechanisms at work in the natural world. Yet the difficulties that attend embracing the theory of natural selection are not felt only by "fundamentalist" Christians or unenlightened persons. Misgivings and misunderstandings regarding evolutionary theory persist, I think, even among those who consider themselves supporters of the theory. The fact that Darwin likened evolution to murder suggests that even he was sometimes uneasy about the implications of the theory that would ultimately bear his name. Nevertheless, murder he did.

Today the theory of evolution plays a surprisingly small role in ethics, and this is all the more extraordinary when we consider how little impact it appears to have had in much of contemporary *environmental* ethics. The arguments that follow grew out of my perception that many ecological theologians have not dealt adequately with the implications of natural selection, despite the relevance of Darwin's theory for their enterprise. The incorporation of the theory of natural selection into environmental ethics is crucial for a number of reasons. In addition to his argument that species are adapted to their environment by means of this process, Darwin also established that species have descended from others, and that all species are related, in varying degrees, to one another. Historically, the recognition that animals and humans are kin gave a greater urgency to the question of how animals ought to be treated—whether animals share our mental, moral, and emotional capacities, whether they have similar "rights," and so on. Aside from issues connected to animal ethics, Darwin's work contributed in other ways to the development of environmental ethics: Darwinism has been a catalyst for the development of a number of ecocentric approaches such as land ethics,[1] and his account of human evolution undermines anthropocentric biases that many theologians and philosophers believe lie at the root of environmental degradation. The theory of evolution has helped to elevate the moral status of all life-forms rather than denigrating the place of humans, defenders of Darwin argue.

Many environmentalists might agree so far with this appraisal of the relevance of Darwin for their enterprise. But in addition to these positive contributions, recognition of natural selection's reliance on processes such as predation, disease, and starvation also serves to highlight a more disturbing picture of the natural world, what Darwin sometimes referred to as the "war of nature."[2] Whether one responds to this view of nature with compassion or revulsion, Darwin's theory brings the issue of natural suffering into sharper focus.[3] Darwin's work also raises more fundamental and, for some, deeply disturbing questions about human nature and the meaning of life on an evolving planet. Darwinism has been blamed for promoting racism and genocide, for endorsing the view that all life is the result of random and purposeless processes, and for undercutting traditional morality.[4] Clearly, Darwin's theory and its implications have been interpreted in a variety of ways.

My primary interest is the relevance of evolutionary theory for environmental ethics and for our understanding of the human-nature relationship.[5] Many environmentalists fail to appreciate its relevance, asserting a view of nature that remains largely untouched by evolutionary theory. This neglect of natural selection produces theoretical and practical problems that will be at the center of this work. I argue that there is a tendency, especially among some Christian environmentalists, to invoke a model of nature as a harmonious, interconnected, and interdependent community. This "ecological model," as it is often called, resonates more with pre-Darwinian, non-Darwinian, and Romantic views of nature than it does with evolutionary accounts. The ecological model, which pervades much of ecotheology, is offered as an alternative to "mechanistic" (or "Newtonian" or "Cartesian") perspectives that regard nature and animals as mere matter and therefore perpetuate dualistic, objectifying, and instrumentalizing patterns of thought and behavior. As a corrective to mechanistic views, the ecological model champions the radical relationality and interdependence of all life. Relationality, in turn, implies an ethic of mutuality, care, liberation, and even love for all other beings, human and nonhuman alike. Ecotheologians discussed here differ in the degree of emphasis they place on each of these ethics (love, liberation, etc.), but on the whole they assent to the ecological model.

Many environmentalists argue that nature presents us with a model and this model has normative import for all our relationships. The concept of nature's interdependence—both in terms of genealogical evolutionary interdependence and interdependence in terms of ecosystemic interconnections and interactions—is central to this model. Among Christian environmentalists the ecological ethic that corresponds to this model is understood to be consistent in important ways with Christianity's ethic of love and care for the neighbor—particularly the neighbor who is suffering, oppressed, and in need, as our "natural" neighbors appear to be. Ecotheologians thus claim that they have grounded their ethics in religious teaching *as well as* scientific knowledge about the natural environment. It is important to keep this claim in mind: my focus will not be on the arguments of creationists or biblical literalists who reject Darwinian theory. Rather, this project examines the work of religious scholars and academic theologians who, *by their own accounts*, have fashioned an environmental ethic compatible with current biological science. I also want to clarify at the outset that, despite the textual and interpretive nature of their enterprise, I assume that ecotheologians do not intend their accounts of the natural world to be understood purely in poetic or metaphorical terms. Indeed, their works contain emphatic denials that they are speaking only on this level.

In the criticisms that follow I am assuming that ecotheologians are serious when they claim—as many of them do—to have grappled with empirical evidence about the origins of life, the evolutionary emergence of humans, the relations of humans to nonhuman life, and the role that natural processes play in the world around us. In the same spirit of seriousness the questions I wish to raise about ecotheological views of nature are shaped by a realistic hermeneutic, an assumption that we can know certain things about the natural world and that such knowledge should direct our treatment of nonhuman forms of life—whether or not we find that knowledge palatable or comforting. However, I am not a biologist or an ecologist. I am acutely aware that I do not possess a thorough, detailed, up-to-the-minute grasp of ecology and evolutionary biology, nor do I expect that anyone who wants to enter into debates about environmental ethics ought to. Ecologists sometimes object to intrusions into their discussions from amateurs and nonscientists; their profession is not a spectator sport.[6] Perhaps not, but this claim is disheartening to anyone who cares deeply about environmental issues but lacks an advanced degree in science. More so than any other science, the concerns of ecology are, or ought to be, the concerns of everyone on earth. None of us can know everything about everything—I take this as both a religious and a scientific truth; realism and humility go hand in hand. But those of us who wish to talk intelligently about the ethical implications of both science and religion have the particular (and not altogether unpleasant) task of trying to come to terms with more information than our "home" disciplines require us to know. My hope is that the present work will bring ecological theology more in line with evolutionary science as it is currently understood.

My focus is largely on the ethical perspectives of a prominent group of ecologically minded theologians that includes Sallie McFague, Rosemary Radford Ruether, Charles Birch, John Cobb, Jürgen Moltmann, Michael Northcott, Larry Rasmussen, and James Gustafson—each of whom claims to have taken natural science seriously in constructing theological ethics that pertain to nature. As I will argue, only Gustafson has adequately met this challenge. I will also deal, in a more limited way, with the work of two secular ethicists: Tom Regan and Peter Singer. The neglect of the natural sciences in environmental ethics is most apparent in Christian ecological theology, but secular environmental ethics often exhibits the same deficiencies, especially in arguments that involve animal liberation and animal rights. Like ecotheology, secular environmental ethics also adopts an inappropriately anthropocentric perspective on animals and nature, failing to draw important ethical distinctions between different kinds of animals. The similar shortcomings in secular arguments will therefore form a secondary topic of discussion in this project.

Each of these authors has made significant contributions to theology and/or environmental ethics; their works are standard fare in many environmental ethics courses. Ecotheologians have listened to critics who charge Christianity with neglecting issues pertaining to nonhuman life and the environment, and their works represent a laudable trend in theology toward discussion of more "earthly" concerns. My study of these theologians examines a representative—but by no means exhaustive—sample of Christian environmentalism. I have included here a few distinct "types" of ecotheology: eschatological, feminist and/or metaphorical theology, as well as process- and liberation-oriented perspectives. Nevertheless, the theologians discussed here are predominantly Protestant (Ruether's approach is more ecumenical and, in places, more sacramental than the others). Though undoubtedly too simplistic, some explanations might be given for the strong presence of

Protestant theology in environmental ethics. One is that Catholic theology, with its emphasis on natural law, has historically allowed for a more normative role for natural processes than Protestant sects have; consequently, Catholics have had far less direct conflict with evolutionary theory and have not undertaken a conscious rapprochement as Protestants have. A second—related—reason may be that Protestantism has been more closely wedded to scripture and its significance for orthodoxy than has Catholicism. As such, Protestants have also been an easier target for critics who trace environmental destruction to certain key scriptural passages that seem to encapsulate and perpetuate the "Christian" anthropocentric, instrumental attitude toward nonhuman life.

As I have already suggested, ecotheologians are often particularly concerned to address the oppression of nonhuman forms of life—including the physical suffering and other bodily needs of individual animals. This focus is consistent with some aspects of Christian teaching. However, as I argue, it is not necessarily consistent with an evolutionary perspective. When proponents of ecotheology address issues of pain and suffering in nature, these issues are often grafted onto a Christian eschatology or liberation theology that is ill-fitted to a serious understanding of the processes of natural selection. When applied to the natural world, some ecotheologians' quest for liberation and healing of the oppressed in nature reflects a persistent reluctance to accept the disequilibrium, moral ambiguity, and ineradicable suffering and death that natural selection entails. Moreover, aspects of evolutionary theory are often employed in a very selective fashion in ecotheology (as well as in secular environmental ethics, about which I will say more in a moment). Biological continuity (genealogical interdependence), for example, is understood to be a key insight of evolutionary perspectives, yet the values and ethics that emerge from this insight remain at odds with an evolutionary/ecological view of nature insofar as the products of evolution are valued far more than the processes. Individual organisms are often the locus of environmental value in these arguments, whereas the context in which those organisms exist may be given insufficient attention.

This is not to say that ecotheologians care nothing about the relations between organisms; on the contrary, the ecological model emphatically endorses a caring, "community" (sometimes organismal) ethic of interdependence. Ecotheologians are clearly concerned with communal dimensions of nature, yet this model retains an inordinate focus on assuring the well-being of each individual creature within the natural community. In human communities this focus is appropriate, but in nature there are no such guarantees of individual well-being. Here, again, I am assuming that ecotheologians' community model—while a model—is intended to contain a kernel of realism: if the details of the model are wrong, the ethic that emerges will, accordingly, be inappropriate. Many such details have to do with the issue of suffering in the natural world. The "problem" of suffering to which I allude throughout this work is, in reality, not just one problem but a cluster of problems. "Suffering," too, has variable meanings in environmental literature, and I will also use this term in close connection with other terms such as *struggle, conflict, strife, predation,* and *competition.*

Suffering is problematic for ecotheology in various ways. In one sense, suffering—as a natural phenomenon rather than an anthropogenically introduced one—is often downplayed or ignored in ecotheology. Environmentalists frequently overlook the suffering and conflict that are an inherent and necessary part of even "healthy" and well-functioning biotic systems. The interdependence of the ecological model is sometimes interpreted as a guarantee of well-being, an antidote to struggle and suffering, for members of the ecological community. This, I will argue, is one problem connected to the issue of suffering in nature: ecotheologians do not pay enough attention to suffering in the allegedly scientific model of nature they construct. While it is true that human activity is the major cause of habitat destruction and species extinction,[7] ecotheologians fail to distinguish these anthropogenic effects from the conflict and suffering that are always and everywhere present in the natural world.

When suffering *is* acknowledged to be part of nature rather than something introduced by humans, it is often regarded as a symptom of nature's fallenness or brokenness—a condition of sin originally introduced by humans' fall from grace.[8] According to this account, nature as a whole is awaiting restoration to its perfect, original condition. In this sense natural suffering is *mistakenly* understood to be a "problem." Part of the problem of suffering, in other words, lies in ecotheologians' belief that it is a problem. By implication, natural selection—which contributes to suffering—is also interpreted to be a problematic and temporary condition of oppressed and as-yet-unredeemed nature. With this interpretation in mind, some ecotheologians encourage us to remove the causes of suffering, wherever and however they occur, while we await nature's restoration. Because this ethic demands our present participation in a process of redemption that can only be completed eschatologically, it is based less on current biological realities than on future hopes and expectations for the removal of suffering. If the ecological model pays insufficient attention to suffering and conflict, the ethic that emerges from this model at times pays *too much* (or the wrong kind of) attention to it, in its quest for the eradication of suffering. In short, we find a lot of confusion surrounding the issue of suffering, regarding how much suffering is actually present, what its causes are, and what human beings can and ought to do about it.

Extending an ethic of love to nature is one means of responding to the problem of suffering and oppressed nature. Loving nature in this manner often means, as it does for theologians such as Sallie McFague, loving in a non-instrumental and unconditional way, recognizing nature's "otherness" in a "subject-subjects" relationship. Loving nature is not just an attitude we adopt, though it is also this. Love requires action. Christians are to extend a loving "praxis" to nature, "treating the natural world in the same way we treat, or should treat, God and other people. . . . We are to love them disinterestedly, for their own sakes, not for ours."[9] McFague sees this environmental ethic as distinctly (if not uniquely) Christian: a love ethic extended to nature is "commensurate with the radical, destabilizing, inclusive love of Jesus."[10] I argue that a love ethic defined in this way is incompatible with inferences from natural science and inappropriate as an environmental ethic.

In the concluding chapters I develop the contours of an alternative environmental ethic that draws upon the theocentric perspective of James Gustafson as well as arguments from land ethics. The perspective of land ethics, first proposed by Aldo Leopold in his classic *Sand County Almanac*, takes natural selection as a guide to environmental policy. Leopold's vision of the biotic community differs significantly from the concept of community that tends to dominate contemporary ecological theology. Leopold and subsequent land ethicists, such as J. Baird Callicott and Holmes Rolston, have stressed that our moral obligations to biotic communities are not analogous to obligations regarding *human* communities, even while humans are "citizens" of the land community. Land ethics does not take humans—or their particular kinds of concerns—to be the center of value. Rather, the centrality of trophic interactions—predator/prey interactions, for example—suggests that life in natural communities revolves largely, though not exclusively, around relationships of struggle, pain, and death. This is not to say that life in biotic communities consist *only* of relationships of conflict, for Leopold's ethic affirmed, and was devoted to preserving, the integrity, stability, and beauty of biotic systems. Yet he recognized that a focus on the larger whole, the ecological and evolutionary context, places restrictions on interventions in nature that would preserve the lives or relieve the suffering of *individuals* within that community. The complexity of this community likewise implies a need for humility and restraint in human interference. In land ethics there is a presumption in favor of what is "natural" to an ecosystem: a native species, for example, is generally given priority over an introduced species, even when granting priority to the former entails suffering or death to other individual organisms within the community.

Holistic land ethics (and particularly some of Holmes Rolston's work) asks us to reflect upon the same sorts of questions and difficulties raised by Gustafson's theocentric perspective. It is not enough that we aim to do what is "good" simply in terms of a human perspective on what is usually deemed good, nor can we assume that doing good all around is possible; we must push further and consider more seriously the questions Gustafson poses: Good for *whom*? Good for *what*? Insights from ecocentric and theocentric frameworks suggest certain conditions of finitude that humans must accept. We are participants in natural processes that we did not create and that do not necessarily conform to human moral preferences and expectations. Such insights may appear simple or obvious to some, yet I would argue that most ecotheologians have failed to grasp them.

In light of this understanding of our role vis-à-vis nature, I argue that a love ethic as articulated by ecotheologians assumes a role for humans that is too interventionist and anthropocentric. Love must not be understood as an imperative to intervene in natural processes in order to eliminate suffering or resolve conflicts. Furthermore, attempts to extend a unified love ethic to humans and all of nature overlook important differences between humans and other animals. Yet a workable environmental ethic should recognize that, as Darwin often emphasized, the differences between humans and other animals are a matter of degree rather than kind. Our ethical relationships with nature are shaped by this constant tension between sameness and otherness. In developing a constructive alternative to current ecotheology, this is one of the parameters that must be kept in mind. To be sure, bonds of love between humans and other animals can—and do—exist, but we should be more circumspect in our attempts to extend a single ethic of love to all living things. A love ethic toward nature must, in other words, be discriminating.

This work examines perspectives from both religious and (to a lesser extent) secular environmental ethics. But I believe the neglect of natural selection in *religious* environmental ethics is especially troubling when the Darwinian account of nature raises certain issues—questions regarding the origin and meaning of suffering and evil in the world, for example—that theology claims to examine earnestly. To ignore Darwinism in ecological theology is a serious oversight. Simply to extend an existing paradigm of love and liberation to the realm of nature, however, is also inadequate.

CHAPTER 1

This View of Life

The Significance of Evolutionary Theory for Environmental Ethics

Nothing is easier to admit in words than the truth of the universal struggle for life, or more difficult—at least I have found it so—than constantly to bear this conclusion in mind. Yet unless it be thoroughly engrained in the mind, the whole economy of nature, with every fact on distribution, rarity, abundance, extinction, and variation, will be dimly seen or quite misunderstood.

—Charles Darwin, *The Origin of Species*

Evolutionary biologist Stephen Jay Gould has written that the "stumbling block" to widespread acceptance of Darwinism lies less in comprehending the scientific details of the theory than it does in the radical message of Darwinian science, namely, "its challenge to a set of entrenched Western attitudes that we are not yet ready to abandon."[1] What exactly is this message? A significant part of it consists in the revelation that humans are not the center of creation—a message that is as simple as it is difficult to grasp. Another disturbing feature of the Darwinian message is that nature operates according to processes that seem wasteful and cruel, mechanisms that cannot easily be attributed to a benevolent creator, that defy explanation in terms of intelligent design.[2] Struggle and suffering are integral to evolution by natural selection—a point that even Darwin found difficult to keep firmly in mind.

The lack of design in nature is a favorite theme of some evolutionary biologists: the best evidence for evolution is not the perfectly formed eye or wing but the parts that are useless, odd, clumsy, and incongruous, such as rudimentary organs that serve no present function. Perfect adaptation is a better argument for creationism than evolution, which takes the circuitous route to adaptation, imperfectly modifying the existing parts and leaving in its path the "senseless signs of history" that are the hallmark of natural selection.[3]

Darwin's theory suggests a distinctive perspective on the world. There are elements of Darwinism that many people have perceived to be disconcerting, both today and in Darwin's time. Yet his theory also laid the groundwork for a number of positive contributions to our view of life and, especially, to our understanding and appreciation of nature and animals. Both the positive and negative dimensions of Darwinism are important for environmental ethics, yet both have been largely ignored or misunderstood by many ecological theologians.

The neglect of evolution appears to be unintentional; ecological theology has attempted, at least since the late 1960s and seventies, to find common ground between evolutionary perspectives and Christian theology. Nevertheless, much of ecological theology holds to an understanding of nature that resembles pre- and non-Darwinian views: often, what I will describe as the darker side of Darwin's theory is downplayed or omitted, and the resulting environmental ethic is inconsistent with well-established knowledge about the natural world. As we will see, ecological theology tends to give priority to the concept of *ecology*—and a particular interpretation of ecology—rather than *evolution*. The ecological model frequently shuts out important elements of evolutionary processes, especially those that seem to contradict or otherwise detract from the ethic ecotheologians seek to derive from nature. This neglect of evolution, and the preference for the term *ecology*, is common in environmental ethics as a whole.

Before turning to a discussion of these two competing paradigms of evolution and ecology, some preliminary points must first be clarified: first, what sorts of ethical considerations are included within the scope of *environmental* ethics? How broadly is this term to be understood? Second, how do contemporary biologists define "Darwinian" evolution and what are some of the points of disagreement among prominent evolutionary biologists?

"ENVIRONMENTAL ETHICS"

The category *environmental ethics* suggests a more coherent and systematic set of issues than is actually the case within the field. In reality, the term covers a wide range of issues that may or may not deserve to be treated as a whole, depending on whom you ask. Environmental ethics usually includes such topics as wildlife management, concerns over deforestation, global warming, loss of biodiversity, overpopulation, and, in some cases, the treatment of farm and laboratory animals (i.e., "nonwild" animals). Humans are often depicted as both perpetrators and the victims of environmental degradation. Many environmental ethicists believe that certain sectors of the human population—especially minorities and women—suffer, along with nature, at the hands of traditionally powerful and privileged classes of people. Thus calls for social justice and environmental healing are issued in tandem by ecofeminists and others who seek to eradicate deeply entrenched structures of oppression and environmental discrimination.

Moreover, many authors address environmental problems from the standpoint of religion; indeed, the environmental crisis is to some *essentially* a religious issue. Many ecological theologians fall within this category. Others, most notably Lynn White, have placed the blame for environmental destruction squarely on the shoulders of religion. Whether or not White's characterization of Christianity as the most anthropocentric and environmentally destructive religion in the world is accurate, it is true that his criticisms forced many Christians to take a closer look at their assumptions about nature.[4]

Probably one of the clearest lines of disagreement within environmental ethics is drawn between those who include the moral status and suffering of individual, nonwild (domesticated, farm, and laboratory) animals within the province of environmental ethics and those who favor a more holistic and ecocentric perspective.[5] For instance, there is often a sharp division between animal activists who focus on the rights or liberation of individual animals and the more ecologically oriented approaches that aim at the preservation of a larger whole, such as a species or an ecosystem. In wildlife management the integrity of an ecosystem may be given priority over the welfare of individual animals. Indeed, overrepresented animals are sometimes killed for the perceived greater good of the ecosystem, or in order to sustain the remaining members of a threatened species. To many animal activists such action may appear cruel and senseless, while to holistic environmentalists the focus of some animal advocates on the suffering of individual animals seems misguided and sentimental. The tensions between some animal activists and holistic environmentalists are heightened by the almost exclusive focus of the former on the issue of animal pain and oppression. Holistic environmentalists point out that pain is not only necessary but even beneficial to the continuing survival and evolution of animal species. Due to their skepticism that a neo-Benthamite ethic revolving around the eradication of animal pain can be extended to wild animals, some environmentalists have denied that the animal rights agenda should in any way be considered a topic for environmental ethics.[6] The split between an ecological ethic and an animal rights position has yet to be completely bridged.

Despite what I think are good arguments for the exclusion of animal rights and liberation from an environmentalist agenda, I will for the sake of argument assume that both holistic and individualist approaches legitimately fall under the heading of environmental ethics.[7] As I will argue in detail at a later point, it is generally inappropriate to extend the same ethic to both wild and nonwild animals. Furthermore, I believe there is considerable confusion in the environmental literature—both secular and religious—surrounding the issue of animal pain as an evil to be eradicated, wherever and however it occurs. However, for now, I will assume that the plight of domesticated animals makes up one subset of environmental issues. In short, in the arguments that follow I will use the term *environmental ethics* to denote ethical arguments regarding sentient beings—both wild and nonwild animals—as well as the ethics for nonsentient nature such as "land ethics."

Amid the diversity of opinions surrounding the definition of environmental ethics, we encounter a set of recurring themes and perennial debates. Chief among these: To what extent are humans really part of nature? Should environmental ethics be ecocentric or anthropocentric (or both at once)? Does an environmental ethic that elevates the moral status of animals dislodge traditional morality and, if so, does it threaten to devalue human life? Or are there already resources within inherited theological and ethical perspectives from which we can fashion an environmental ethic? Many of the fundamental challenges posed by the enterprise of environmental ethics—questions about the role of humans in nature and the limitations of traditional religion and morality—intersect those raised by an evolutionary perspective. We will return to this cluster of complex issues frequently in the arguments that follow. But first we must consider the question of what constitutes an "evolutionary perspective" and examine some of the issues about which "Darwinists" disagree.

EVOLUTION WARS

Throughout this work I often refer to Darwinism or evolutionary theory as the dominant paradigm for understanding the natural world and human-animal relationships. In reality, however, there is no single, undisputed definition of these terms; many who consider themselves Darwinists disagree about the details of evolutionary processes. In order to understand current debates among biologists regarding these details, it is important to keep in mind that Darwin himself had no viable account of the mechanisms of inheritance. Only relatively recently, when the theory of natural selection was synthesized with Mendel's research on genes (in the 1940s), did the bigger picture of evolution by natural selection begin to take shape. By the mid-twentieth century the "modern synthesis" in biology had successfully incorporated Mendelian genetics and Darwinian natural selection; the resulting theory became known as neo-Darwinism. Debates continue to this day regarding such topics as the role of genes in evolution, the rate at which evolution occurs, the significance of the selection process in producing new forms, and the moral implications of evolutionary theory for human society. Much of this discussion has taken place among several prominent biologists, many of them affiliated with Harvard University, including Stephen Jay Gould, Richard Lewontin, Richard Levins, Edward O. Wilson, and British biologist Richard Dawkins.

Let us look first at the issue of rates and patterns in evolutionary processes. Stephen Jay Gould notes that the "gradualist" account (evolution by slow, steady change), which has been the conventional interpretation, may not be entirely accurate. Much of Gould's work is critical of what he calls "extrapolationism" in evolutionary biology.[8] Extrapolationism, in Gould's view, is the fallacy of making assumptions about the overall history of life (that is, evolution on a very large scale) based on shorter-term, local changes within species. Gould argues that biologists erroneously assume that large-scale evolutionary change is simply the cumulative result of local, minor changes. In place of gradualism Gould (along with Niles Eldredge) has proposed a theory of "punctuated equilibrium" that posits evolution by fits and starts, periods of rapid speciation followed by times of relative stasis. Evolutionary stasis does not mean that no change whatsoever takes place between one generation and the next. Rather, these small changes are expressed as minor deviations from and fluctuations around a "phenotypic mean."[9] For example, climatic changes may increase selection pressure for a particular beak size or shape, but ultimately, over long stretches of time, the change is transient and nondirectional (not leading to a new species, for instance). These small-scale, local deviations can be extrapolated (within reason) to large-scale relative stasis that is reflected in the fossil record. But while the explanatory move from local to large-scale seems valid enough with regard to static periods, such extrapolation is questionable for the process of speciation (evolution during periods of nonstasis). Gould denies that microevolutionary Darwinism operating at the level of individuals within populations can account for the more dramatic, macroevolutionary changes that have occurred in the history of life.[10]

Apparent gaps in the fossil record may be more consistent with Gould's theory of evolution than they are with gradualism, which has more difficulty accounting for intermediate, "transitional" forms. Owing to Gould's attention to such gaps, he has often been an unwitting tool in creationist attempts to dismantle evolutionary theory. Because Gould appears to be "admitting" that evolutionary biology is in trouble—that the gaps in the fossil record are real, prevalent, and problematic for the dominant paradigm of gradual Darwinian evolution—his work is (mistakenly) interpreted by some as bolstering an antievolutionary agenda. Of course, in Gould's view the gaps are not necessarily gaps or missing links at all; rather, they reflect the reality of periods of relative inactivity in evolution. Moreover, the "rapid" changes that Gould attempts to explain are rapid in geological terms. His views support neither a young earth account of life nor a belief in the sudden creation of distinct, fully formed species.

Other contemporary debates have to do with currents in sociobiology and disagreements regarding the unit of natural selection. While Gould understands natural selection as a process taking place among whole organisms or groups of organisms such as species (a process known as "species selection" or "species sorting"), some biologists, including Richard Dawkins (*The Selfish Gene*) argue that genes and gene lineages are themselves engaged in evolutionary competition. In Dawkins's account, genes, rather than organisms, are fit or unfit; strong selection pressures are exerted on genes primarily and on their bearers secondarily. Of course, Dawkins and Gould agree that evolutionary change and genetic change are related. What they disagree about is how simple and straightforward that relationship is. Some biologists understand Dawkins to endorse an overly simplistic, direct, deterministic relationship between genes and traits; genes, and gene expression, they insist, are multifaceted and thus the relationship is far more complex than Dawkins concedes. But from the standpoint of gene-centered evolutionists such as Dawkins and E. O. Wilson (discussed in more detail below), the critics have both exaggerated the amount of complexity in the gene-trait relationship, and misrepresented their views as overly deterministic.[11]

In light of disagreements about the role of the gene as the unit of selection, it is clear why biologists also debate the importance or unimportance of natural selection vis-à-vis other factors such as variation. Gould, for example, places less emphasis on selection in determining the direction of evolution than does Dawkins. The power of selection, Gould maintains, is always limited insofar as it can only work on the raw material that is present. The existing gene pool of variations and the conditions imposed by the evolutionary past of populations restrain the power of natural selection to produce change. Gould, along with another Harvard biologist, Richard Lewontin, argues that evolutionary biology has at times been too "adaptationist." That is, the discipline as a whole has been preoccupied with the search for adaptive fitness in evolution and scientists have been too willing to assume that 1. natural selection modifies organisms toward some particular end and 2. that biologists can discover with some certainty what that end is.[12] Both Dawkins and Daniel Dennett (*Darwin's Dangerous Idea*) have been the target of the Gould-Lewontin antiadaptationist campaign. Gould and Lewontin charge Dawkins and Dennett with generating evolutionary "just-so" stories—panglossian arguments that attempt to explain virtually every trait in terms of adaptive function.[13] Gould urges scientists to pay more attention to the nonadaptive cases in evolutionary biology, such as odd or imperfectly formed structures, and argues for a multilevel analysis of evolution that encompasses genes, individuals, and species.

Of course, many of the issues debated by contemporary Darwinists—particularly those revolving around the role of genes in determining traits and behavior—have larger social and political implications; indeed, scientists' political leanings may significantly influence the kind of research agenda that emerges. In this context, probably no topic in modern biology has generated as much controversy and heated exchange as sociobiology, first introduced by Harvard biologist E. O. Wilson in 1975.[14] Sociobiology, and its offshoot, evolutionary psychology, examine the biological bases of social behavior.[15] Wilson's sociobiological perspective has been challenged, even condemned, by *left*-leaning biologists such as Lewontin, Gould, and Levins (see Levins and Lewontin, *The Dialectical Biologist*). Wilson shares Dawkins's focus on low levels of organization such as genes and, according to critics, both biologists fall into the hyperadaptationist/genetic determinist camp. Opponents of sociobiology may understand selection as operating at higher levels and, as noted above, are often skeptical of causal connections between genes and behavior.

Sociobiology emerged in part as an explanation for certain vexing problems in Darwinian theory, especially the existence of altruistic behavior in animals. In a conventional, individualist account of natural selection, it is difficult to understand why one organism would risk its life for others (as when animals issue warning calls upon spotting a predator in their midst). Such behavior appears maladaptive, since altruistic organisms take on increased risks to themselves and may not survive to pass on their genes. One camp known as group selectionists attempts to explain altruistic phenomena by shifting the focus to a higher level, viewing the group rather than the individual as the vehicle for natural selection. According to this analysis, more cooperative groups are more likely to survive *as a whole*. As plausible as this may sound, however, critics (including sociobiologists) have pointed out that an individual may always "defect" or "cheat" on this arrangement, fail to reciprocate altruism, thereby gaining an advantage and passing on his own genes. In other words, group selectionists must explain the fact that selection at the individual level and selection at the group level seem to work against one another. This problem does not plague species selection (Gould's theory) for the simple reason that altruistic traits such as warning calls or food sharing may belong to individuals as well as groups, whereas properties that Gould has in mind—geographic range, for example—are *emergent* properties, belonging only to the species.[16] Thus in species selection, higher- and lower-level selection are not in tension. Sociobiologists, as well as many other biologists, reject group selection,[17] offering alternative, more "selfish" explanations for apparently altruistic behavior. One of these is the idea of kin selection wherein organisms aid their genetic relatives in order to pass on some of their own genes (which of course are shared by the relative). Sociobiology contends that ostensibly altruistic behavior is in reality a strategy with beneficial consequences accruing for the individual (or more accurately, *via* the individual) at the genetic level.

In *Sociobiology* Wilson—whose research is primarily in entomology—extended such analyses to human society, arguing that genes play an important role in shaping human social and sexual behavior. His claims were met with a severe backlash from scientific peers, particularly Lewontin who, along with Gould and other members of a Harvard sociobiology study group, associated Wilson's work with racism, genetic determinism, and crass reductionism. For his own part, Wilson claims to have been a political naïf who never glimpsed the controversial social and political ramifications of his work. In the late 1970s the sociobiology controversy culminated infamously in an incident that took place at the American Association for the Advancement of Science, in which Wilson's detractors chanted protests and even poured a pitcher of ice water over his head.[18] Wilson and Lewontin have been accusing each other of allowing ideological commitments to distort science ever since. We will revisit some of these debates, particularly those surrounding genetic determinism and Wilson's sociobiological claims in later chapters.

What do these varying definitions of Darwinism imply for the scientific critique of ecotheology and environmental ethics that I present here? Is it anachronistic in light of developments in modern biology to define Darwinism in terms of suffering, struggle, and conflict between organisms? If adaptationists are right about the tendency of evolution to produce fitness, does this fly in the face of Darwinism as competitive and profligate? Not necessarily. Despite disagreements about the details of evolution, few scientists would deny that suffering and struggle play an important role in evolution. One need not hold to a view of evolution as perpetual, unmitigated struggle and strife in order to see that some ecotheological arguments have little or no grounding in evolutionary biology and ecology. Even gene selectionists/adaptationists such as Dawkins understand evolution as an unseen war played out between gene lineages striving to replicate. Organisms (or survival machines, as Dawkins sometimes calls them) are the medium in and through which the rival genes compete. Dawkins's adaptationism (if that is what it is) in no way vindicates pre-Darwinian natural theology; nor does his work lend credibility to modern design arguments, such as those proposed by religiously minded intelligent design theorists (indeed, intelligent designs theorists such as Michael Behe identify Dawkins as an opponent rather than an ally). Dawkins himself is an avid atheist who believes that Darwinism supports his position.

For biologists such as Gould, on the other hand, struggle and competition are features of evolution as well, but these take place between individuals or, more likely, between entire species, since emergent properties such as geographic range and total gene pool variation belong to species *as a whole*, not merely to individuals (and certainly not to the gene). Again, the disagreement is about the level at which Darwinian struggle unfolds. It is worth noting, too, that debates regarding the unit of evolution have undergone rapprochement in recent years, as scientists have begun to argue that gene selection is not necessarily at odds with selection at higher levels. It may be that evolutionary struggle occurs simultaneously at multiple levels, though it is not clear that terms such as *struggle* are appropriate when applied to genes. Darwin himself argues in *The Origin of Species* for such an expansive notion of struggle, which includes interactions between organisms and abiotic aspects of their environment; struggle, in other words, did not necessarily refer to individual organisms engaging in direct "battle."

Disagreements among Darwinists, and departures from Darwin's original formulation of natural selection, do not indicate that the theory is in disarray or that "Darwinism" has ceased to have any coherent meaning. Even Gould, whose work significantly modifies Darwin's understanding of such important features as the patterns, rates, and units of selection, denies that his account invokes mechanisms not found in the original theory. In the opening pages of his last major work, *The Structure of Evolutionary Theory*, Gould argues that Darwin's theory retains a "'happy' intellectual status" in modern science, with "enough stability for coherence and with enough change to keep any keen mind in a perpetual mode of search and challenge."[19]

Which of these Darwinists presents the darker vision of evolution? The answer depends largely on your perspective. Despite his opponent Dawkins's evident fondness for depicting gene lineages engaged in survival strategies (Dawkins likes to portray evolution as "red in tooth and claw"), Gould's vision of nature might be construed as more disturbing, violent, and morally ambiguous, on the whole, than gene theorists.' Not only is he far less convinced of the prevalence of adaptation in evolution; his version of Darwinism also stresses contingency and more numerous, widespread, and sudden extinctions (Gould's book *Wonderful Life*, in particular, emphasizes the theme of contingency). He is skeptical—far more so than Dawkins or Wilson—that evolution is a progressive process, steadily producing greater and greater fitness in organisms. Gould is highly (and in my view appropriately) critical of biologists, past and present, who have attempted to base sweeping claims about human nature on biological data. Critiques of such misuse of science are a recurring theme in Gould's voluminous writings, most notably in *The Mismeasure of Man*.

On nearly all these issues, to the extent that I can follow the details of the debates, I see far more merit in the Gould/Lewontin/Levins position (I critically examine some of Wilson's views, for instance, in chapter 6). Gould and Lewontin may overstate the extent to which rival biologists view nature through a hyperadaptationist or panglossian lens. But such critiques certainly apply to some ecotheological discussions of nature in which cooperation, perfect adaptation, symbiosis, and mutuality are said to prevail. I am not denying (nor would most evolutionists, for that matter) that there are instances of adaptive fit, reciprocity, cooperation, etc., in nature, but I am contending that some ecotheologians and environmentalists emphasize these features disproportionately. The current state of evolutionary science—despite a lack of consensus among Darwinists—supports the contention that these features represent only one side of the story. A number of themes that emerge from debates among modern Darwinists will be relevant to the discussions that follow. These include the importance of struggle in nature and the levels at which it operates, how prominent a role natural selection plays and how widespread adaptation is in the natural world, and the question whether or not evolution shows any discernible direction, especially when that direction is taken to be morally significant. Perhaps the most important question is whether individuals—as opposed to a collective entity such as species, populations, or biotic communities—are or ought to be the unit of moral consideration in environmental ethics.

Let us consider the rest of the story, the part that is downplayed in some environmental and ecotheological literature. The following discussion will focus particular attention on the terms *evolution* and *ecology*, examining some of the history of these concepts and considering the ways in which these terms are used, if they are used, by religious and secular environmentalists.

EVOLUTION AND ECOLOGY

Contemporary ecotheologians are particularly fond of the ecological model of nature. But what is the relationship between this model and an evolutionary account? When an environmental writer such as Sallie McFague sets out explicitly to adopt an "evolutionary, ecological" perspective on nature, is it legitimate to use *evolution* and *ecology* in conjunction and interchangeably, or do these terms signal distinct models for understanding the natural world? The term *ecology* did not appear until the nineteenth century, 1866 to be exact, when it was coined by biologist Ernst Haeckel. Today *ecology* conjures up for many people a model of nature as an interconnected and fairly stable system of complex relationships. Often the *ecosystem* concept plays a central role in discussions of ecology. I will use this term guardedly, acknowledging that certain features of the definition, as well as concepts associated with ecosystems such as "community," remain problematic.

Frequently embedded in the concept of an ecosystem is the assumption of an intricate and often precarious community of living things, a balance of nature that is increasingly threatened by human encroachment and disregard for the environment. Serious disturbances to this system, typically brought on by humans, can upset this balance; if disturbance is serious enough, the entire structure may collapse, perhaps taking us along with it. Environmentalists, both religious and secular, often exhort us to live within the limits set by nature, to help restore nature to its proper, ecological conditions. Unlike many other sciences, ecology has enjoyed a great deal of popular enthusiasm, especially since the 1960s. The ecologist is the most beloved of scientists, the scientist whose noble task it is to protect nature and its inhabitants.[20] For many people today, Donald Worster argues, "ecology has come to represent the arcadian mood that would return man to a garden of natural peace and piety." The ecological vision, at least in the popular imagination, is often one in which nature's "most fearsome aspects have been shut out."[21]

Evolution, as we have seen already, may suggest precisely those fearsome aspects: a fierce, competitive struggle for food, shelter, and mates. The processes of evolution are described with such words as *blind*, *random*, and *purposeless*. Natural selection, we are told, operates according to "chance"; in fact, some recent evolutionists have begun to highlight the aspects of evolution that reveal its deeper connections to chaos theory. This picture of nature is one that is already as disturbed as it is disturbing: perhaps there is no "balance" in nature, no homeostasis to be disrupted. The processes of evolution appear unpredictable and wasteful; if organisms "relate" to one another at all, they do so primarily in terms of eating and being eaten. On this interpretation nature appears to be a system that is not especially conducive to sustaining harmonious relationships with humans or any other creature.

So, is nature harmoniously balanced and stable or do disequilibrium and conflict reign? This brief sketch of both models of nature is, of course, overly simplistic. Yet the history of ecological science and evolutionary theory is in many ways the history of a contest between two competing paradigms. If it is possible, as I believe it is, to draw ethical implications from the natural world and human origins, then it matters a great deal which account of nature we put the most stake in. Ecotheologians have put the most stake in an account that has little to do with Darwinian evolutionary theory.

PRE-DARWINIAN ECOLOGICAL PERSPECTIVES

The term *ecology* is closely related to the term *economy*. Both words are derived from the Greek *oikos* meaning "household" and, in the eighteenth century, the arrangement of nature was understood to be something very much akin to an orderly household. Eighteenth-century views of nature frequently revolved around the concept of an economy of nature, a benevolent arrangement in which all living things are "so connected, so chained together, that they all aim at the same end and to this end a vast number of intermediate ends are subservient."[22] Myriad examples of the fitness that exists between an organism and its environment were taken as proof of God's design, expressed in the organization of nature. Before concepts of evolutionary ancestry, organisms were classified according to structural similarities to one another, "elements of a divine plan" reflected throughout the natural world.[23] The work of the eighteenth-century taxonomist Carolus Linnaeus was central to this understanding of nature. Linnaeus's great contribution to biology, the well-known system of binomial nomenclature of each plant and animal (the labeling of each as belonging to a particular species and genera) embodied the notion of nature's economy. His work assumed without question that nature was "designed by Providence to maximize production and efficiency."[24] The entire structure was held together by relationships of benevolent interdependence and mutual assistance. Organisms at or near the top of the economy were thought to be dependent on those beneath in order to survive, and each creature was believed to occupy a precise place, its own office, with its own resources, within the larger economy. The interdependent structure of the system ensured that no creature would suffer from a scarcity of food or the effects of overpopulation. Nature's beneficent arrangement, revealed in the Linnaean system of classification, was a testimony to God himself, who "ensured the permanent stability of the system by designing a series of checks and balances that would maintain the population of each species at an appropriate level."[25]

The economy of nature is one important strand within eighteenth-century science. Another related current of thought sprang from the Romantic tradition. Romantic naturalism is often portrayed as a reaction to the Enlightenment utilitarianism of the economy of nature, yet the two perspectives are similar in some respects. The Linnaean economy had definite mechanistic overtones—nature was portrayed as a "well-oiled machine."[26] This account of nature conveniently resembled the factory system that would soon dominate English urban existence. In contrast to this perception of nature as fairly static machine, the Romantics drew inspiration from a world that was very much alive and changeable, a natural world envisioned as a single living organism. Both models, however, relied heavily on notions of the interdependence and relationality of all beings in nature.

In Romantic ecology each part of the system was thought to be crucial to the survival of the whole, as in the Linnaean model, but the analogy with a machine composed of interchangeable parts was consciously rejected in favor of a communal, organismal metaphor. The Romantics viewed nature as constituted by "inviolable interdependence . . . a system of necessary relationships that cannot be disturbed in even the most inconspicuous way without changing, perhaps destroying, the equilibrium of the whole."[27] Our illusion of separation and independence from nature is sustained with great cost to ourselves and other organisms, this view suggests. Humans, as inherently part of this system and yet alienated from it, should seek a new, more intimate relationship to nature as the only cure for spiritual and physical wounds. Nature was understood as the macrocosm mirroring the human microcosm, a place that both nurtured and reflected human relationships, values, and states of mind. "The key word in the Romantic vocabulary was 'community,'" Worster argues, and, especially, a *community of love*, an "extended net of natural relationships."[28] In short, the Romantic view amounted to what many environmentalists now regard as an ecological perspective.

Recent ecological models, including those depicting ecosystems as holistic, integrated, interdependent communities, clearly owe much to Romantic thinking. Contemporary environmental ethics rooted in this model of an ecological community often exhorts us, as the Romantics also did, to repair and restore our severed connection to the natural world as a way of healing both ourselves and the environment. For example, in a work entitled *Nature's Web* Peter Marshall depicts modern ecology as centered on the idea of nature as a "unified and balanced organism." Ecology, he argues, has firmly established that "all life is interdependent, and that the earth itself is a self-regulating organism."[29] Marshall understands the Romantic movement to have "paved the way for an evolutionary and ecological appreciation of nature," despite the fact that evolution plays little or no role in this account.

Concepts of nature as a self-regulating organism, or a community of benevolent arrangements revolving around some principle of interdependence, have obvious appeal, vague though they may be. As ecologist Frank Benjamin Golley observes, such concepts have long provided humans with a sense that "somewhere out there, there was ultimate order, balance, equilibrium, and a rational and logical system of relations."[30] Many writers in the field of environmental ethics, both religious and secular, have upheld an image of nature as a community founded on interdependence as the key insight of ecology and have, consequently, interpreted nature's interdependence and unity normatively.[31] The community ethic implied by the ecological model is one we should all live by, McFague argues: "The ethic that emerges from the ecological model is care for all those in the community."[32] Like the Romantics, she calls for an ethic of "loving nature." McFague tempers this optimistic assertion somewhat with the qualification that "an ethic of community is not all love and harmony," but she concludes that an ethic rooted in the ecological model demands, at the very least, respecting the "otherness" of all living things within the community. Committing ourselves to the good of the community (understood in its broadest, ecological sense) repairs the damage we have done to nature and enriches our lives by bringing us back into the relationship: "The ecological [model] . . . suggests the renewal of a subject-subjects relation."[33] McFague, like Marshall and many other environmentalists, believes that such an ethic is rooted in an "*evolutionary, ecological*," understanding of nature.[34] Yet in this account, as in Marshall's, little attempt is made to distinguish the terms *evolution* and *ecology*; rather evolution is subsumed under a normative, community definition of ecology.

ECOSYSTEM CONCEPTS

But are such interpretations of nature and ethics consistent with modern evolutionary and ecological science? Romanticized notions of natural communities as self-regulating, goal-seeking superorganisms persist in much of environmental ethics, even while they are regarded as "anathema to most ecologists."[35]

The superorganism and community metaphors for nature predate the term *ecosystem*, yet these ideas are frequently, and mistakenly, conflated. The ecosystem idea was first proposed by British ecologist Arthur George Tansley (1871–1955) in 1935. American ecologist Frederic Clements (1874–1945), on the other hand, was a chief proponent of the superorganism view of nature, though the concept of nature as a single, living organism has much older roots. Clements recognized that the superorganism changed over time, but he "abhorred chaos in ecological thought," arguing that the superorganism showed steady movement (referred to as ecological succession) toward a relatively permanent, harmonious, and stable "climax community."[36] The climax community could persist indefinitely, barring significant disruption from natural disasters, such as forest fires, or anthropogenic sources of damage to nature. Clements's concept of nature as a single, organic whole lends itself to an ethic of preservation—an ethic that typically blames human interference for disruptions and imbalances in nature. As such, it has great appeal for environmentalists to this day.

According to Golley, Tansley formulated the ecosystem concept in contrast to the superorganism and community metaphors that had long pervaded natural science.[37] Since metaphors for nature were (and are) often the product of thinking analogically from human communities to natural "communities," they were at times imbued with "positive, almost idyllic" meaning.[38] More specifically, as far as Tansley was concerned, Clements's ecology took the organism and succession metaphors too far and exaggerated the holistic, collective action of natural systems. Tansley attempted to analyze natural systems as wholes, without resorting to organism or community metaphors that were potentially misleading in their neglect of the relationship between organisms and the surrounding physical and chemical environment. Tansley's ecosystem concept was offered as a means of incorporating the organic and the inorganic factors, a bridge between biotic organisms and components of the environment that were typically not included as living "members" of the community.[39] Organismic and community metaphors, in other words, paid insufficient attention to crucial but "nonliving" aspects of nature.

Tansley's ecosystem was comprised of a number of "subordinate parts," which included "energy flows, biogeochemical cycles, communities, and populations of species."[40] Ecosystem ecology has retained this hierarchical ordering for many years. A "community" here (and in the modern sense) is not the same thing as an ecosystem but generally refers to one level within the system: groups of species populations living in a specific region. Yet ever since Tansley's formulation, many people (including ecologists) have continued to associate ecosystems with the older concepts that Tansley sought to supplement or correct: "Ecologists tended to misuse the term ecosystem as a more modern expression for the community concept or Clementsian complex organism and thus maintained the confusion that Tansley was trying to overcome."[41] Thus Golley contends that some modern ecologists have thrown out the ecosystem idea owing to a *mistaken* interpretation of ecosystems as a "continuation" of the superorganism and other ecological concepts whose time has passed.[42]

There is an ongoing debate among ecologists regarding the definition and usefulness of concepts such as the ecosystem. Whereas Golley defends the concept—shorn of certain associations such as the self-regulating superorganism idea—others, such as Joel Hagen, point to tensions, even incompatibilities, between the ecosystem idea and the field of evolutionary ecology. I will examine these debates in more detail when we consider some recent developments in ecology and evolutionary science. For now, the key points are that 1. models of nature that pervade much of ecotheology and environmental ethics gloss over evolutionary processes and interactions among organisms, focusing almost exclusively on broader ecological themes and 2. those ecological themes—particularly superorganismic, mutualistic, community themes—are themselves considered outdated, inadequate, perhaps entirely obsolete, by many scientists. My disagreement, however, is not with the fact that some environmentalists are guilty of deriving normative guidelines from nature—an "ought" from an "is"—but rather that the *ought* they are deriving represents only a part of nature's *is*, as science understands it.[43] Darwinism, properly understood and incorporated, offers a corrective to this particular type of myopia in environmental ethics. Before turning to more recent discussions of the relationship between evolution and ecology, let us step back for a moment and consider some of the features of Darwin's theory that distinguished it from earlier conceptions of nature.

THE DARWINIAN PERSPECTIVE

Darwin's theory, like the Romantic tradition that is still evident in some ecological concepts, also offers an account of the *relatedness* of all life. In fact, evolutionary theory strengthened the notion of human kinship with animals by explaining its true biological, not just metaphorical, significance. But, more important, what Darwinism contributed to the study of nature was a clear counterpoint to the impression of a benevolent, communal interdependence in nature. Again, this is not to say that a Darwinian view consists solely of competition and struggle;[44] nor were earlier accounts of nature such as those developed by Linnaeus and the Romantics oblivious to nature's darker side. But before Darwin elements of conflict and disorder were often deemphasized or rationalized as aiming toward some eventual, teleological good. "That the natural economy had its interdependent social ranks and classes appealed to naturalists as a firm guarantee against Hobbesian violence," Worster argues, however there remained a "residuum of suspicion that perhaps Hobbes was right about nature after all."[45] With the advent of Darwinism the suspicion that the natural world was the arena of a war of all against all appeared to be confirmed. Let us look more closely at these disturbing features of evolution as Darwin understood it—features that ecological theology often sidesteps.

The Dark Vision of Darwinism

In his travels abroad during his *Beagle* voyage, Darwin was greatly impressed with the instability, competitiveness, and violence of the natural world. Like many of his age, he had been steeped in the Romantic tradition, and much of his writing reflects the aesthetics of that movement. But Darwin returned from his travels armed with volumes of data that would cast doubt on this view of the natural world. In his own plodding and meticulous fashion, Darwin would eventually call into question not only earlier, Romantic views of nature but also many of the central tenets of traditional theism.

Although the characterization of evolutionary theory as "survival of the fittest" does not quite capture the complexity of the theory,[46] it is certainly true that the notion of struggle among nature's creatures was a catalyst for the development of Darwin's theory, and many modern Darwinians likewise affirm that struggle and competition remain important. The impact on Darwin of reading Malthus's 1838 *Essay on the Principle of Population* has been documented repeatedly by historians of science. Darwin's insight was that the struggle for limited resources raging in human society was occurring in the natural world as well. Malthus had argued that human population increases geometrically while the resources necessary for life increase only arithmetically. Food supply in particular would always lag behind. Welfare measures would only lead to increases in population and thus "liberal-minded efforts to better society must inevitably bring nearer the misery, starvation or war by which the excess population would eventually be eliminated."[47] Only moral restraint, Malthus suggested, could prevent this stark scenario.

Malthus's account struck a chord with what Darwin had already observed in the natural world. He began to wonder how the variation that existed in nature might be related to the outcome of the struggle for existence. Commenting on the impact that reading Malthus's essay had on the development of his theory, Darwin wrote in his autobiography, "Being well prepared to appreciate the struggle for existence which everywhere goes on from long-continued observation of the habits of animals and plants, it at once struck me that under these circumstances favourable variations would tend to be preserved, and unfavourable ones destroyed."[48] What seemed to be a designed, perfect fit between an organism and its environment was in fact the outcome of chance—chance in the sense that those favorable outcomes are the result of processes not aimed at their production.[49] An organism adapted to its environment is the product of a number of factors, many of which were not pleasant to contemplate: overpopulation, predation, competition, disease, and starvation. Nature no longer appeared to be a place of balance, a single self-regulating organism. At best, balance and self-regulation could only be achieved at the expense of many animal lives and as the net result of great suffering. Whatever balance might be achieved through natural selection seemed short-lived. At the same time that Darwin was pondering the struggle for existence in nature, the fossil record had begun to suggest that mass extinctions were frequent and widespread phenomena throughout the earth's history. By the end of the nineteenth century the natural world as depicted by Darwinian theory had undergone a hideous alteration.

When, after a lengthy delay, Darwin finally published his theory in 1859, he described a process that results in the survival and reproductive success of individuals or groups best adapted to their environment, leading to the perpetuation of qualities suited to that particular environment. As articulated by Darwin, the basic features of natural selection are 1. individuals in a population vary (i.e., genetically—though, again, Darwin had no modern concept of the gene), 2. organisms produce more offspring than will survive to maturity, and 3. in competition for limited resources individuals with an advantageous trait are more likely to survive and to leave more descendants than those lacking it. These favorable traits therefore accumulate in the population because of natural selection.

The genetic material from which natural selection "selects" exists randomly.[50] That is, it results from random variation due to genetic recombination or mutation. It is these elements of chance and randomness of natural selection that sometimes lead people to understand evolution as a blind, mechanistic process, a force in nature with no agency—or at least none that seems related to the will of a benevolent deity. Darwin's theory is often characterized as Cartesian: a lawlike system of matter in motion, without meaning or purpose, the origin and functioning of which can be explained without any recourse to a creator. On this account Darwin's theory seems to reaffirm the exact picture of nature the Romantics had earlier rejected.

The evolutionary model unveiled by Darwin depicted a world not only much starker than earlier views but also much more ancient than anyone had previously believed. Darwin essentially "historicized" science. In *The Discovery of Time* Stephen Toulmin has argued that Darwin's theory

> called into question both the uniqueness of man, and the traditional view of cosmic history. Being an *historical* theory, Darwin's explanation of the origin of species inevitably challenged the historical claims of all alternative, religious accounts of the Creation of Man. . . . Man's relationship to God in History was the central theme of the Christian message, and no

> theologian could happily see all of God's works turned into by-products of the blind action of natural selection.[51]

Darwin recognized more than anyone the uneasiness his theory might generate. He sensed the burden of the facts he had uncovered and, in particular, he was aware that natural selection highlighted the suffering in nature that had lurked beneath the surface in Romantic ecology. As far as undermining theology was concerned, Darwin professes to have shed his religious beliefs rather painlessly. "The very old argument from the existence of suffering against the existence of an intelligent first cause seems to me a strong one."[52] In his autobiography Darwin recounts his struggle to come to terms with what theologians call the problem of evil. For him the problem was especially intractable in the animal world, more so than with human beings for whom suffering might at least "serve for [our] moral improvement."[53] James Rachels points to Darwin's struggle with issues of theodicy in order to emphasize a crucial point about the evolutionary perspective—namely, that Darwin unearthed a genuinely difficult challenge to traditional theism, one that theology has tried ever since to bury. The suffering of animals in nature, in other words, is really *the* problem for theodicy:

> Like many doubters before him, Darwin appealed to the existence of evil as a reason for rejecting theism. But he strengthened the argument by giving it two distinctive twists. First, Darwin realized that the traditional debate centres entirely on evils related to *human* life and history. The evils that need justifying have to do with *man's* suffering, and the traditional theistic justifications have to do with *man's* comprehension. . . . But typically, the first thing that occurs to Darwin is that human life and history are only a small part of nature and its history. Countless animals have suffered terribly in the millions of years that preceded the emergence of man, and the traditional theistic rejoinders do not even come close to justifying *that* evil.[54]

Rachels and Toulmin agree that Darwin brought to light new and more troublesome problems for theism. The historicization of science effected by evolutionary theory both broadened and sharpened the focus of the problem of evil. The usual set of responses to this problem—that evil is a consequence of our free will, or a punishment for sin, for instance—tacitly assume the presence of humans from the beginning of time. Yet the "struggle for life," showcased in Darwin's theory, "is pre-human in origin and is built into the very way in which the world is wired for complexification of life."[55] Darwin himself recognized the difficulty of accepting the view of nature his theory presented and he understood, too, that it constituted a break with a more palatable account. It should be noted, however, that Darwin sounds a somewhat Romantic note when he suggests, in the midst of this discussion of animal suffering, that *on the whole*, "according to my judgment happiness [in nature] decidedly prevails, though this would be very difficult to prove."[56] At the very least, he argued, natural selection, unlike theism, made some *sense* of the fact of suffering, even if it could not guarantee compensation for the suffering endured by individual organisms.[57]

Darwin's darker vision of nature has never completely triumphed over a pleasant, harmonious interpretation. In the popular imagination, as well as in the science of ecology, there has been an ongoing tug-of-war between ecological models of harmony and evolutionary accounts of struggle and disorder. The more recent history of concepts of evolution and ecology has important implications for environmental ethics. We will now turn briefly to this recent and quite complicated interplay of ideas.

POST-DARWINIAN ECOLOGY: DISCORD OR HARMONY?

Obviously, the history of evolutionary theory and its relationship with ecology does not end with the publication of Darwin's works. Subsequent history tells of a continuing debate—increasingly, a dialectic—revolving around ecological and evolutionary paradigms. The development of the ecosystem idea that emerged from a critique of Clements's superorganism concept is one example. Within the science of ecology the concept of the ecosystem still has great explanatory power; some ecologists defend the ecosystem model as relevant and useful, once it is properly understood. But in the years since Tansley's use of the term, the ecosystem idea has been the target of much criticism, sometimes from those proposing a more "Darwinian" model.

Early ecosystem models focused attention on the collective level (such as trophic levels or the behavior of aggregates of organisms); natural selection, on the other hand, was understood to operate at the level of the individual organism engaged in competition (assumptions still questioned by some biologists, as we have seen). Ecosystem studies were at their height of popularity in the 1950s and sixties, owing largely to the publication of Eugene Odum's classic textbook on the subject, *Fundamentals of Ecology* (with editions in 1953, 1959, and 1971), but the field was developing with little or no direct reference to evolutionary theory. Odum's concept of the ecosystem stressed the communal benefits of cooperative interaction between the ecosystem's members. Like Clements before him, he believed that ecosystems brought "order and harmony out of the chaotic materials of existence."[58] Competition between individuals also occurred in Odum's system, but he maintained that the survival and homeostasis of the unified ecosystem depended upon counterbalancing competition with cooperation. Group selection and coevolution provided a crucial offset to competition and contributed to the overall stability of ecosystems.[59] Ultimately, Odum's ecosystem concept (which became *the* ecosystem idea, as it is popularly understood today) embraced an overt form of holism that Tansley had set out to critique when he first employed the term ecosystem twenty years earlier. In its emphasis on cooperation and organismlike self-regulation, Odum's ecosystem concept also bears traces of a much older Romantic perspective. Also, like the Romantics, some ecologists in the 1950s and sixties saw their mission as "ecologizing" society as a whole, since nature was perceived as a model for human communities, a corrective to our disruptive tendencies.[60] Ecotheologians have carried on this tradition of adopting nature's ideal community as a blueprint for human society.

By the 1970s the case for group selection had been largely discredited, leaving ecosystem ecologists with the difficult task of explaining the relationship between individual competition and overall communal stability. In other words, the assumed counterbalances to competition—such as coevolution, altruism, and group selection—no longer appeared entirely defensible. Without these, how could *de*stabilizing processes such as individual selection and competition produce and maintain homeostasis and balance in the ecosystem? Perhaps the belief in balance was itself suspect.

Meanwhile, evolutionary science had undergone radical changes of its own, owing to the modern synthesis of the 1940s that united genetics with other features of natural selection. But these changes were perceived as largely irrelevant to ecosystem studies. By the 1960s, Hagen argues, there was a "growing belief among many younger ecologists that evolutionary theory had been ignored or misunderstood by their elders."[61] Evolution and ecology were developing in relatively separate spheres, and their separation eventually led to a clash between ecosystem studies and evolutionary studies.[62] The divergence of ecosystem and evolutionary studies, moreover, was occurring at a time of heightened interest in environmental problems, and the public was turning to ecologists for answers. Ecological and evolutionary scientists were unable to present a unified front. Finally, the centennial of Darwin's publication of *The Origin of Species* brought with it a renewed interest in evolutionary theory and forced ecosystem scientists to reexamine the impact of evolution on "the development of behavior and life history, the regulation of populations, and the organization of communities."[63] A new field of evolutionary ecology had emerged, but it was not entirely clear that ecosystem studies could be integrated with the new discipline.

Challenges to ecology from evolutionary science did not stop there. The late twentieth century saw the rise of "nonequilibrium ecology," as some ecologists began to argue that computer and mathematical analyses (analyses of gene frequencies over time, for instance) revealed a fundamental unpredictability in the natural world; ecological changes appeared to occur at random.[64] By the 1990s one leading ecology textbook devoted only a single paragraph out of six hundred pages to the ecosystem concept—a dramatic change from the ecosystem-centered ecology of the fifties and sixties, as exemplified by Odum's classic text.[65] Ecology was "perceptibly moving away from that unified theory that had sought to bring the living and nonliving together into a single, coherent, balanced, and orderly system."[66] The extreme randomness of nature, some ecologists have suggested, is consistent with Darwinism; in fact, some have begun to understand the processes of nature as an illustration of chaos theory. Such challenges to ecology from Darwinism are ongoing and the ecosystem concept "continues to come under attack from critics for its lack of evolutionary emphasis."[67]

If the chaos challenge proves to be correct, ecologists may be forced to conclude that they can never accurately describe a biotic system's state at any point in time, nor can they predict future events—the likelihood of survival of its species, for instance. From an environmentalist's perspective, this trend in ecology is an alarming one. As Daniel Botkin observes, modern ecology has opened up a Pandora's box of resource management issues. When applied to real environmental cases—to marine fisheries, wildlife, and endangered species, for instance—concepts of nature's balance or stability have proven to be deficient and ecology, it seems, has been forced to jettison its once central tenets. What then becomes of resource management and environmental conservation? The discovery of disorder, of continual fluctuation, as something intrinsic to nature makes resource management difficult to defend, much less implement. Once we concede that some changes in nature are "natural," good, or valuable, how can we then "argue against any alteration of the environment?"[68] If nature is so fundamentally erratic, so naturally inclined to disorder and disturbance, how can we know what our impact on the environment is? Which disturbances in nature are caused by us and which are inherent in the system? How can we even judge that "disturbance" is a bad thing in the first place? Furthermore, how can we know that our efforts to cultivate environmental responsibility, to preserve biodiversity, have any positive results without any background stability against which to compare our efforts? This vision of nature provides no norm whatsoever. As a basis for environmental ethics, this brand of "Darwinian" ecology seems to counsel doing nothing.

Would a thorough incorporation of a modern evolutionary perspective lead, ironically, to a complete split between evolutionary science and the goals of environmentalists? Are proponents of chaos ecology right to see Darwin as their champion? I do not find this argument wholly plausible. Darwin himself certainly did not understand nature to be completely devoid of order, and, though he stopped well short of Romantic appeals to the harmony of nature, he often envisioned the competitive struggle as leading to some more favorable outcome.[69] Those who equate Darwinism with chaos are overstating the case as much as those who interpret it as perfect adaptation and order. Yet one effect of the contest between the ecosystem paradigm and that of Darwinian science has been a reconsideration of the role of competitive struggle in biology. In the 1950s and sixties the rediscovery of Darwin in modern ecology meant that ecosystem scientists could no longer discount the role of natural selection within ecosystems. As it stands today, the science of ecology is endeavoring to incorporate elements of order and disorder in nature, competition and cooperation, parts and wholes. Even proponents of disequilibrium ecology often perceive a certain regularity to the chaotic shifts in natural systems. It may also be that fluctuations at one level of an ecosystem can prevent fluctuations at higher levels, creating a buffer for the ecosystem as a whole. Likewise, those who defend the ecosystem as a meaningful (and, to some extent orderly) model of nature acknowledge that concepts such as homeostasis, self-regulation, and equilibrium must be qualified in light of scientific developments. Take the concept of equilibrium, for example: Golley argues that equilibrium in ecosystems is not so much a condition of "balance" or "stasis" to which a system naturally returns as it is a "response system, that is in a dynamic relation with its environment."[70] According to this "relativistic theory of ecosystem behavior," equilibrium is better understood as a statement regarding the history of a system, an account of its "past performance" rather than a "prediction about its future state."[71] Though ecologists may never predict the behavior of an ecosystem with complete accuracy, he argues, they can "describe broad limits of possibility."[72]

Where does this leave us? Many ecologists have begun to regard order and chaos as part of the same picture, "two distinct manifestations of an underlying determinism."[73] Nature is more correctly understood as complex rather than thoroughly chaotic. What this sketch of the Darwinian versus ecological (or eco*system*) paradigms should caution environmentalists to remember is that the ecological themes of the balance of nature, the harmoniously interdependent nature of life, are not only too vague but too one-dimensional as well. They continue to capture only one aspect of nature. As Botkin observes, the "true idea of harmony of nature" as understood in modern ecology, "is by its very essence discordant, created from simultaneous movements of many tones, the combination of many processes flowing at the same time along various scales, leading not to a simple melody but to a symphony at some times harsh and at some times pleasing."[74] While they may disagree on the extent to which older views of ecology emphasized balance and the extent to which recent models truly point to chaos, ecologists, and historians of ecology, generally agree that there has been a movement away from cooperation, coevolution, harmony, and balance as the predominant ecological themes. Yet, many environmentalists, particularly ecotheologians, still invoke an allegedly naturalistic, scientific norm rooted in these ideas. It may not be an overstatement to say that Christian interpretations of nature have never completely come to terms with the evolutionary perspective. But why is this? Is Darwinism such a dismal outlook that its incorporation into environmental ethics has proved too depressing and onerous a task? Part of the explanation may lie in the complex history of evolution and ecology that I have just sketched. Given that ecosystem ecology is not consistently well-grounded in evolutionary science, it is not surprising that ecological theologians neglect evolution as well.

Or perhaps there are other, more insidious, factors at work. Does the omission of Darwinism reflect a suspicion that the theory of natural selection threatens to destroy traditional tenets of theology and theodicy? We have seen that Darwin himself perceived his theory as a challenge to basic Christian beliefs. Perhaps he was right. Before turning to a discussion of the positive contributions of Darwinism—the "brighter side" of the evolutionary perspective—let us briefly consider some explanations for why environmental ethics (and especially ecotheology) lags behind biological science.

WHY THE GAP?

The gap between environmental ethics and evolution may have more to do with the methods of scientists than with the fears of theologians. The notion that moral implications, good or bad, can be derived from nature has been forcefully discredited by some of the staunchest proponents of evolutionary theory. Rachels points out that "friends" of Darwin may very well hesitate to promote the idea that Darwinism poses a threat to religion and values, even if they believe this to be true. After all, the perception of such a threat is what has often kept Darwinism out of public schools, thanks to the right-wing "enemies" of Darwinism. Admitting that Darwinism constitutes a distinct set of values, a sort of "worldview" in and of itself, plays directly into the hands of opponents of evolution who charge that the theory is really a form of quasi-religious dogma—secular humanism—masquerading as objective science in biology text books.

Hoping to sidestep charges of dogmatism, evolutionary scientists themselves have lent credibility to the notion that Darwinism, as a set of "facts," can have no moral implications. Stephen Jay Gould insists that facts of nature can pose no threat to "moral values." Science, he argues "can no more answer the questions of how we ought to live than religion can decree the age of the earth."[75] Science and religion exist in separate spheres, what Gould calls nonoverlapping magisteria (NOMA): "Science gets the age of rocks, and religion the rock of ages; science studies how the heavens go, religion how to go to heaven."[76] Since, according to this view, no conflict can exist between science and religion, why should theologians trouble themselves with learning the finer points (or even the broad contours) of scientific theories? If ethics and theology lack a firm grasp of science, scientists themselves may be partly responsible for perpetuating the gap between theology and science. Whether or not evolutionary scientists truly believe that there are no moral implications of Darwinism, they may adopt that position, if only to protect the theory from assaults from without.

It is difficult to banish the specter raised by Darwin himself that evolutionary theory *does in fact* pose a threat to traditional theology, theodicy, and morality and that Christianity cannot truly absorb the shock of evolution *and* keep traditional theism intact. This is the position Rachels, for example, has taken, arguing unequivocally that the theory of evolution "undermines theism."[77] "In particular, it undermines the traditional idea that human life has a special, unique worth."[78] Rachels maintains that evolution has forced a gradual "retreat" of traditional theism, persistently edging it out until finally we (like Darwin) must ask ourselves: "If religious belief is reduced to this, is it worth having? What remains is a 'God' so abstract, so unconnected with the world, that there is little left in which to believe."[79] The best we can hold onto, Rachels argues, is some fragile strand of deism. The very concept of God loses its meaning in light of evolutionary theory:

> There is now far less *content* to the idea of God. The concept of God as a loving, all-powerful person, who created us, who has a plan for us, who issues commandments, and who is ready to receive us into Heaven, is a substantial concept, rich in meaning and significance for human life. But if we take away all this, and leave only the idea of an original cause, it is questionable whether the same word should even be used.[80]

Even if Darwinism does not "make theism impossible," he concludes, it "still makes it far less attractive than ever before."

If Rachels is correct, perhaps ecological theology and much of theological ethics as a whole has ignored Darwin because it is simply not possible to maintain any robust system of theology in light of evolution. Ecological theologians are forced into professing a more sanguine, pseudo-Darwinian view because the alternative would be to abandon any recognizable form of theism altogether.

Rachels's argument involves questionable assumptions, however: his contention would seem to be that once we jettison the belief that humans were specially created by God, with a unique and privileged status (he refers to this tenet as the "image of God thesis"), traditional theology collapses altogether. He (mistakenly) rejects the possibility, in other words, that anything like traditional theistic belief can endure without an essential core of anthropocentrism. Rachels argues (correctly, I think) that Darwinism discredits the "image of God thesis"; in his view, traditional theism, minus the image of God thesis, equals deism—the belief that everything was created by a being who can in no way be assumed to have an ongoing, personal relationship to what has been created. If we agree with Rachels, then a true "ecological theology," one that does more than pay lip service to evolution, is an impossibility, an oxymoron.

Rachels's argument is difficult to dismiss but not impossible to refute. James Gustafson's theocentric ethics and Holmes Rolston's biological and theological arguments offer a powerful rejoinder to Rachels's claims. The issue whether or not it is possible to retain traditional theological orientations—in light of evolutionary theory and without the core of anthropocentrism—is one that will receive fuller attention in subsequent chapters. The focus of the remainder of this chapter will be some of the more *positive* contributions of Darwin's work to environmental ethics. Having considered some of the reasons that a Darwinian perspective might be difficult to adopt, we should also give some attention to the constructive role Darwin's work has played in the development of environmental ethics. I will offer these positive contributions without critique; later we will examine these contributions in greater detail and, especially, the claim that evolutionary continuity implies equal moral consideration for all living things. At present my intention is to suggest that the significance of evolutionary theory for environmental ethics makes the absence of a Darwinian perspective all the more puzzling.

THE BRIGHTER SIDE OF DARWINISM: POSITIVE CONTRIBUTIONS TO ENVIRONMENTAL ETHICS

If the Darwinian view served to swing the pendulum toward a darker, more disturbing vision of nature, historically the theory of evolution also led many Victorians seriously to reconsider their relationship—literally—with animals. In Darwin's time the "dismal science" of evolution (as Worster calls it) also fostered a more ethical treatment of animals and greater interest in and concern for nature as a whole.

There are at least three implications of natural selection, or "Darwinism" generally, that are directly relevant to the development of environmental ethics: animal-human kinship, the ethical status of nonsentient nature, and the issue of anthropocentrism.

Animal Ethics

Although animal protection and nature preservation movements predate Darwinian theory, Darwin's work galvanized the efforts of early environmental groups and lent additional weight to their arguments. The recognition that humans and animals are biologically related further reinforced the moral imperative to treat animals humanely. In the nineteenth century "animal protectors naturally followed the progress of Darwinism with keener than ordinary interest," James Turner has argued. "For them evolution had become a simple matter of fact—no longer an issue for debate but an assumed starting point for further discussion of the relations of people to animals."[81] Indeed, the Victorian era has been characterized as a time of heightened awareness of pain and suffering; natural selection extended that concern to include the suffering of nonhuman animals. Certain developments in the nineteenth century were crucial for impressing upon the Anglo-American mind a concern for animal welfare:

> One was the realization that human beings are not supranatural but are directly descended from beasts. The other was the rising esteem for science, as a model of intellectual endeavor and as the key to the future of the race. It is not so widely recognized, though much has been written around the edges of this subject, that the nineteenth century was also an era of enhanced sensitiveness about pain.[82]

Although Darwinism has often been portrayed as a Cartesian perspective, in one very important respect Darwin's work is quite explicitly anti-Cartesian. Darwin's study of animal emotions and states of mind in *The Expression of the Emotions in Man and Animals* reads as a direct rebuttal to Descartes's claim that animals are mere machines, automata possessing no rational faculties. It is striking that in their persistent efforts to dislodge a Cartesian framework ecotheologians rarely if ever make use of Darwin's animal studies (though secular animal advocates such as Regan and Singer have done more with this). Darwin argued repeatedly that animals exhibit a wide range of complex emotions and rational faculties. Descartes had admitted that animals possessed something like emotions, but he likened these responses to mechanical reactions, devoid of any reflective element. A tortured animal's cries of pain were akin to the ringing of a bell—involuntary, automatic responses involving no thought processes.

Darwin's extensive writing on animals paints a very different picture of them as our kin who share many of our most basic emotions and social behaviors. *The Expression of the Emotions in Man and Animals* is filled with sketches of animals displaying a wide range of facial expressions—smiling, frowning, gestures of disappointment, sulking, terror, and wonder. Most notably, Darwin also attributed to them what he called "intellectual emotions," behaviors combining complex mental and emotional responses that differed from humans' only in degree, if at all. He recounts experiments, both his own and those of other researchers, that seem to prove beyond any doubt that animals employ reason and reflection, that they are able to recall past events and deliberately plan future actions.

In the nineteenth century scientists who experimented on animals, especially vivisectionists, invoked Cartesian arguments about animals' lack of mental awareness and inability to process pain. With the publication of Darwin's works, antivivisectionists recognized him as an important and powerful ally and were eager to see evolutionary theory accepted, if only for the sake of animals. Darwin himself was not so sure whose side he was on. Despite his concern for animal suffering—he pronounced vivisection for the sake of mere curiosity "damnable and detestable"—he was uneasy about being asked to take a stand against the methods and discoveries of other scientists.[83] He was convinced that the science of physiology had benefited greatly from animal experimentation, though he confessed that the mere thought of vivisection rendered him insomniac. He always argued for anesthesia during vivisection—anesthesia was not widely employed for most of the nineteenth century—and he once lamented to his daughter that he felt the pain of the word *vivisection* so keenly that he wished "some new word like anaes-section would be invented."[84]

Anecdotes of Darwin's kindness to animals abound. He stopped strangers in the street to chastise them for whipping their horses; the nineteenth-century antivivisectionist Francis Power Cobbe described him as a man who "would not allow a fly to bite a pony's neck."[85] Antivivisectionists saw Darwin's reluctance to take up their cause as a kind of weakness of character, a streak of hypocrisy. Yet Darwin's tolerance toward animal research stemmed in part from his hope that research involving vivisection would ultimately produce discoveries to *control* suffering and disease. Darwin "would sanction even pain to put pain to flight."[86]

But, for the most part, Darwin refused to get embroiled in this debate, as he refused to join in many other controversies to which he was asked to lend his considerable scientific authority. It would appear that he took some of the implications of his work for animal welfare less to heart than did many of his advocates. Yet his own research on animals—which never involved vivisection—certainly advanced the cause of animal ethics. Despite his personal ambivalence on the details of their treatment, he did much to elevate their moral status. Darwin remains a pivotal figure in the movement toward a more ecocentric view.

Land Ethics and Moral Extensionism

If a greater concern for our animal relations was one of the contributions of Darwinism to environmental ethics, a second, more indirect, effect was the extension of moral consideration to all nature, including, ultimately, the land itself. Just as Darwin's account of physical evolution underscored the biological continuity of humans and animals, his theory of *moral* evolution identified a concern for all living things as the pinnacle of ethical development. Roderick Nash observes in *The Rights of Nature* that environmental awareness can be traced largely to Victorian thinkers and especially to the ethical implications of Darwin's work in *The Origin of Species* (1859) and *The Descent of Man* (1871). Darwin's theory influenced American environmentalists as well. Nash points out that Aldo Leopold's formulation of land ethics in the 1940s relied heavily on Darwin's account of the evolution of ethics in *The Descent of Man*; Leopold's treatment of ethics in his much celebrated *Sand County Almanac* "nearly plagiarized Darwin."[87]

Darwin envisioned an evolutionary process involving an ever expanding circle of moral concern for others that would eventually embrace all creatures on earth. Moral consideration for animals, Darwin argued, was the highest attainment of ethical evolution, the human species' "latest moral acquisition."[88] Darwin and Leopold agree that a lack of concern for other living things reveals a deficient, primitive moral development. The sportsman who never learns to regard animals as anything other than "trophies" or "certificates" to capture and display is "the caveman reborn," Leopold argued.[89] Taking his cue from Darwin, Leopold extended consideration beyond concern for animals to the land itself, maintaining that such an extension of ethics "is actually a process in ecological evolution." "There is as yet," Leopold argued, "no ethic dealing with man's relation to land and to the animals and plants which grow upon it." Yet he believed that an ecological conscience would evolve: "The extension of ethics . . . is, if I read the evidence correctly, an evolutionary possibility and an ecological necessity."[90] Current proponents of the land ethic such as J. Baird Callicott and Holmes Rolston continue to make Darwinism and natural selection central to their work. Callicott, for instance, has presented Darwin's account of the evolution of moral sympathies as a foundation for environmental ethics.[91] For the most part, however, most environmentalists—especially Christian environmentalists—have paid little attention to Darwin's theory of natural selection and even less to his account of the evolution of ethics.

The Critique of Anthropocentrism

Clearly, Darwin's work in biology and ethics contributed much to the development of moral concern for nonhuman animals and nature as a whole. Darwin's theory has also done much to undermine cherished anthropocentric notions regarding the status of humans in the natural world. Because environmentalists of all varieties have often identified anthropocentrism as a major contributing factor in our reckless destruction of the environment,[92] the critique of anthropocentrism is a position that environmental ethics would seem to share with a Darwinian view of nature. As philosopher Mary Midgley has tirelessly argued, the critique of anthropocentrism is a *Darwinian* critique, and yet the myths of both science and religion continually marginalize this perspective: "Groundless fantasies about a dazzling human future, both on this planet and off it, are developed to justify our chronic abuse of nature," Midgley writes. "Most of them rest on extrapolating graphs of human development that contain nothing to justify any such extension. Evolution itself is imagined, quite contrary to Darwin, as following such a graph."[93] Despite the widely held assumption that natural selection has permanently dislodged humans from the center of creation, an "apotheosis of man" has persisted, ironically with evolutionary science as its driving force:

> Evolution, in these prophecies, figures as a single, continuous linear process of improvement. In the more modest form in which some biologists have used it, the process was confined to the development of life-forms on this planet. But it is now increasingly often extended to do something much vaster—to cover the whole development of the universe from the Big Bang onward to the end of time—a change of scale that would be quite unthinkable if serious biological notions of evolution were operating.[94]

Darwin's dethronement of humans remains incomplete. Like other environmentalists, Midgley worries about the effect that our myths of human progress and perfection have had, and continue to have, on our natural environment.[95]

Midgley's critique of the persistence of anthropocentrism (or "reductive humanism" as she labels this bias within the sciences) has an interesting counterpart in James Gustafson's criticism of theology. In *Ethics from a Theocentric Perspective* Gustafson observes that "the dominant strand of Western ethics, whether religious or secular, argues that the material considerations for morality are to be derived from purely human points of reference." He proposes an alternative set of questions: "What is good for the whole of creation? What is good not only for man but for the natural world of which man is part? What conduct is right for man not only in relation to other human beings but also in relation to the ordering of the natural and social worlds?"[96] The persistence of the anthropocentric perspective in Western ethics and theology is connected, in Gustafson's view, to the fact that ethics and theology have also remained strangely untouched by developments in the sciences. Science, particularly natural science, lends support to the idea that the world is not made for us, that it is not completely under our control. The recognition that the world is not made by us or for us is an idea that, for Gustafson, is theologically significant: "The experience of the ultimate power and of powers bearing down upon, sustaining, and creating possibilities for action induces or evokes piety."[97] A sense of piety is reinforced by data from the sciences, data from which we can infer that we are merely "participants" in, rather than controllers of, our world: "A proper understanding of ourselves and our participation requires that we have in view a broader context of interactions than is central to some other versions of ethics."[98] Science can help to broaden our view. Thus Gustafson's criticism of anthropocentrism in theology is part and parcel of a criticism of theologians' reluctance to take the implications of science seriously. A more accurate—and more Darwinian—understanding of humans' place in nature challenges anthropocentric ethics that have persisted for so long in the West. This understanding is badly needed in environmental ethics, as it is in ethics and theology generally.

Both the negative and positive features of Darwinism are important for environmental ethics, though neither translates directly or simply into an environmental ethic. Darwinism sheds light on a vision of nature that is, admittedly, more disturbing than earlier views; without this vision the ecological perspective as often articulated by some ecotheologians is incomplete, inadequate, and outdated. At the same time that evolutionary theory poses difficult challenges to theology and ethics, it has also contributed significantly and constructively to increasing our understanding of natural processes, to improving our ethical relationship with other animals, and to clarifying the place of humans within the natural world. For all its discomfiting implications, there is still, as Darwin himself said, grandeur in this view of life.

On the whole, many of the problems besetting environmental ethics and ecological theology stem from an account of nature that remains largely impervious to what Gustafson calls inferences that can *reasonably be drawn* from the sciences. Even in the absence of complete consensus among those who consider themselves Darwinian scientists, we can draw some general and well-founded conclusions about the natural world and its processes. In the discussions of ecological theology that follow, Gustafson's critique of theology's neglect of scientific knowledge will periodically serve as a touchstone in my own criticisms. Although, for the most part, he does not direct his criticisms toward *environmental* ethics and theology, his arguments are especially trenchant in these disciplines, where scientific grounding is so clearly warranted.

In the next two chapters we will turn to a critical examination of the current state of ecological theology, focusing upon the works of Sallie McFague, Rosemary Radford Ruether, and Michael Northcott (chapter 2) and Jurgen Moltmann, Charles Birch, John Cobb, and Larry Rasmussen (chapter 3). My criticisms of these writers will center on the absence of a Darwinian perspective, the preference for the "ecological" over the evolutionary (illustrated in the ecological model), and the issue of suffering in nature, which is a prominent concern of many ecotheologians.

CHAPTER 2

The Best of All Possible Worlds

Ecofeminist Views of Nature and Ethics

All nature is but art, unknown to thee;
All chance, direction, which thou canst not see;
All discord, harmony not understood;
All partial evil, universal good;
And, spite of pride, in erring reason's spite,
One truth is clear, Whatever is, is right.

—Alexander Pope

Robert Booth Fowler observes that environmentalists frequently turn to the idea that nature provides an ideal model of life in community. For some Christians an ecological ethic requires them to "live the community ethic," represented by nature's "holism, interrelatedness, and the vital importance of striving to maximize a common good."[1] Yet this turn to nature for a model of community life overlooks important questions, including whether natural systems actually exhibit communal, harmonious characteristics. Scientists, he notes, generally see far more "competition and instability in nature than community enthusiasts might like."[2]

The work of two prominent ecofeminists, Rosemary Radford Ruether and Sallie McFague, illustrates this pro-community sentiment particularly well. As we will see, the interpretation of nature as normative for human relationships involves a much greater emphasis on themes of *ecology* than *evolution,* as I delineated these concepts in the previous chapter; often the ecological data invoked is outdated or contains misrepresentations. As such, both Ruether and McFague present arguments that are incompatible with current scientific understandings and both exhibit serious shortcomings as ethics that could realistically be applied to nature. Following a discussion of

Ruether and McFague, we will conclude with an examination of the work of Michael Northcott, who is sympathetic to ecofeminist perspectives regarding the relationality of all life-forms even while he is critical of many ecotheological claims. All three ecotheologians propose an ethic revolving around community/covenant models and all, in constructing an environmental ethic, engage in a selective use of ecological and evolutionary data.

The distinction between an ethic *derived from* nature and one *extended to* nature becomes blurred in the writings of some ecotheologians. Ruether attempts to locate an already existing norm within nature and then encourages a conscious emulation of it in human communities. Yet her reading of nature is one-sided; she isolates "ecological" themes that support the basic ethic of mutuality and interdependence she espouses. In this sense her ecological ethic is as much a wishful imposition on nature as it is a derivation from it. McFague begins, for the most part, with a human (Christian) ethic and argues that it can be extended, without significant modification, to the natural world. Her argument moves back and forth between a Christian and an ecological model, illustrating how they can mutually maintain and derive insights from one another. Both models, she suggests, support an ethic of community, care, and love for others.

As we will see, the relationship between what McFague calls ecological "reality" (i.e., science) and the "Christian subject-subjects model" becomes convoluted and rather confused. Ruether's understanding of the relationship between normative clues in nature and the ethics of human societies is less complex. However, both she and McFague present nature in a non-Darwinian, romanticized light, echoing and affirming older, often pre-Darwinian perspectives, while glossing over elements of what I have described as the components of an evolutionary perspective.

We will begin by looking at Ruether's account of natural communities and human ethics in *Gaia and God*, noting her use of such terms as *ecology*, *interdependence*, and *community*. The portrait of nature that emerges is not entirely inaccurate, but it is clearly one-dimensional in its neglect of evolutionary processes involving conflict, predation, and competition.

RUETHER'S ROMANTIC ECOLOGY

A recurring theme in Ruether's work is the contrast between the harmony and cooperation of natural communities and the discord, competition, and strife of human societies. We should heed the example of nature, she argues, as a model for correcting our own relationships of domination and oppression. Such relationships perpetuate our alienation from other human beings and from nature as a whole. Western consciousness manifests a dangerous split between concepts of male and female, nature and culture, and even between knowledge, on the one hand, and "wonder, reverence, and love" on the other.[3] In our steadfast denial of kinship with all other beings, we have constructed "arrogant barriers" that harm all forms of life, human and nonhuman alike (48). Attending to nature with an eye toward modeling our relationships upon ecological communities can help to bridge the gaps in our own consciousness and in our relations with other beings. "Human ethics," she writes, "should be a more refined and conscious version of this natural interdependency, mandating humans to imagine and feel the suffering of others, and to find ways in which interrelation becomes cooperative and mutually life enhancing for both sides" (57). The "science" of ecology provides us with a "vision of humanity living in community with all its sister and brother beings" (58).

How does Ruether understand this science? In the introduction to *Gaia and God* she begins by defining key terms in her argument including *ecology*, *feminism*, and *ecofeminism*. Ecology, she writes,

> comes from the biological science of natural environmental communities. It examines how these natural communities function to sustain a healthy web of life and how they become disrupted, causing death to plant and animal life. Human intervention is the major cause of such disruption. (1)

Note how Ruether's sense of ecology stresses the view of nature as a "community," a "web of life" in which order and harmony would reign, were it not for human activity, which disrupts this natural system, bringing death to living things. Her definition presupposes that such disruption is foreign to nature, that it is something imposed from without by humans.

Having defined ecology, Ruether proceeds to clarify what earth *healing* entails. The aim of ecofeminism—a perspective on nature generated by the intersection of ecological and feminist concerns—is to bring about a "healed relation" to the earth, to create "just and loving interrelationship between men and women, between races and nations, between groups presently stratified into social classes, manifest in great disparities of access to the means of life" (2–3). Healing the earth can only happen in conjunction with healing our relationships to one another, by eradicating relationships of domination and exploitation and replacing them with loving and just relationships. At one level Ruether's argument revolves primarily around improving relationships between different kinds of people. Taken as a whole, however, her argument assumes three levels of (frequently strained) relationships: those between peoples of different race and gender, that between all people and the earth, and those existing within (nonhuman) nature as a whole. I will have more to say about these categories in a moment. For now, let us see how Ruether understands evolution and what the relationship is between evolutionary processes and the ecological model she espouses.

THE NORMATIVE ROLE OF NATURAL SCIENCE: EVOLUTION AS COOPERATION

Ruether provides a sketch of the history of science and its impact on traditional Christian understandings of the creation of the earth and the place of humans within the rest of nature. She then turns to a discussion of "Nature's Laws and Human Ethics," which contains the core of her argument about the normative import of scientific views of nature for human ethics. Here Ruether frequently falls into what Allen Fitzsimmons calls a "deviation-from-nature standard of measurement," whereby the natural world is judged to be inherently good or moral and human disruption of or deviation from nature's norm is dangerous or sinful.[4] She immediately brings the term *ecology* into the discussion, contrasting its overtones of communal harmony with the human propensity for disruption and domination. "Ecology is the biological science of biotic communities. . . . Its study also suggests guidelines for how humans must learn to live as a sustaining, rather than destructive, member of such biotic communities."[5] She argues that in the case of modern ecology science *appropriately* resumes a role that it once played (usually inappropriately) in the past vis-à-vis religion and morality, namely, "as normative or as ethically prescriptive" (47). The fundamental normative guideline provided by science is the principle of *interdependence*. "One of the most basic 'lessons' of ecology for ethics and spirituality is the interrelation of all things" (48). Apparently, Ruether thinks that evolutionary processes (distinct from this benign reading of ecological interdependence) have no such normative import, or she prefers to avoid the normative implications of evolution.

Ruether does not always invoke terms such as *interdependence* and *interrelationship* as though they are sufficient and self-evident proofs of her normative claims. She subtly defines the notion of human "kinship" with nature in a number of eloquent passages describing the affinity between elements in the human body and the bodies of distant stars and planets. She marvels at biotic cycles that return dead organisms to the soil and infuse new beings with life. Her portrait of the intimate relationship between human life and the rest of the universe is both moving and, essentially, accurate from a scientific standpoint. Yet, on closer inspection, it is an incomplete portrait.

Ruether's account of nature pays particular attention to large-scale, life-sustaining cycles of interdependence such as "food chains," "photosynthesis," and the "cycle of production, consumption, and decomposition" (51). Her examples are chosen with two primary goals in mind: to illustrate the beneficial ordering of nature and to contrast it with the inferior and destructive forms of social and political interaction typical of human communities. Nature is orderly and efficient—the "cycle of production, consumption, and decomposition prevents waste from accumulating"—whereas humans have only recently perceived the wisdom of recycling our waste. Nature provides a system of "feedback" that keeps plant and animal populations balanced; human population, by comparison, represents an anomalous "extreme case" of proliferation of one species that threatens to overrun and destroy all others (54). This account of nature emphasizes overarching themes, patterns, and cycles that provide ethical directives for humans: "Cooperation and interdependency are the primary principles of ecosystems" (56).

Ruether is much less interested in particular encounters between organisms or groups of organisms—those involving struggle, predation, competition—that do not seem to provide evidence of a community spirit and life-enhancing cooperation in nature. Or, more precisely, she simply redefines these encounters as specific instances of the broader ecological themes she approves. The common good of the ecological community is promoted even in these particular cases that appear, at first glance, to run counter to community harmony. When we turn to Ruether's few remarks about evolution, it becomes even clearer that her description of nature shuts out all the potentially unpleasant and (as a norm for humans) ethically problematic aspects of natural systems.

On the rare occasions when Ruether examines evolutionary processes, the word *evolution* is consistently modified so as to deemphasize connotations of competition or conflict. She always interprets evolution as a cooperative venture, culminating in the creation and maintenance of community life. Evolution, in other words, is understood as a process *aimed* at community existence and cooperation. An interesting example of this is that when the term *evolution* appears in *Gaia and God* it is almost always accompanied by the prefix *co* or followed by the words *interdependency* or *interrelationship.* Ruether hopes to impress upon her readers the predominance in nature of "*co*evolution" and "*coevolutionary interdependency*" citing numerous examples of evolution as fostering "basic codependency" in nature" (49–52). A bison will allow a bird to ride on its back because of the services rendered (eating insects that attach to the bison's fur). Animals form cooperative groups that provide "mutual help," such as food sharing, care, and protection for the young (56). Ruether's only brush with a darker vision of Darwinism is her brief acknowledgment in one sentence that "most animals are herbivores, but some eat other animals" (51–53). Yet, depending on how the terms are defined, one could in fact argue that most organisms are *predators*. Heterotrophs (organisms that eat other organisms) far outnumber autotrophs (those that synthesize their own food). The category of predation is not necessarily limited to carnivores, as this category is often understood (e.g., large vertebrates such as cats who stalk their prey), and instances of parasitism, which may also be considered a form of predation, are rampant in nature. In any case, our eyes are quickly averted from this glimpse into predatory behavior, as Ruether assures us that "in nature, death is not an enemy, but a friend of the life process" (ibid.).

Because she wishes to uphold an ecological ethic of interdependency and coevolution, Ruether must dispel any "survival of the fittest" reading that might creep into a nature-informed ethic.[6] She invokes this phrase and expresses concern that such a reading of nature could distort an ecological ethic by mistakenly implying that "the strong have a right to prevail over the weak and that, in the competitive struggle for existence, might makes right" (55). The view of nature as "red in tooth and claw" and the image of the natural world as an arena of "carnivore animals killing other animals" is, Ruether argues, "a vastly distorted picture of nature" that generates a "false ethic of competition" as destruction (55–57). She dismisses this distorting and disturbing vision, arguing that the preoccupation with competition in nature "fails to recognize that all the diverse animal and plant populations in an ecosystem are kept in healthy and life-giving balance by interdependency." She then makes the classic Romantic move of rationalizing apparent competition and conflict as *aiming toward* a higher, purposive end:

> Competition, that is, control of each population in relation to others by a pattern in which some populations feed others, works to sustain the total

> life of a biotic community only because it is set within a larger pattern of mutual limits. (55)

Note how Ruether has carefully defined "competition" so as to highlight its *community*-building aspects. Competition, like evolution, becomes "*interdependent* competition": "There are many patterns of cooperation, as well as interdependent competition," she stresses (56). Having defined ecology as the science of biotic communities, characterized by coevolutionary processes and competitive interdependency, she puts all the pieces together and presents us with a scientific, normative ecological ethic:

> Cooperation and interdependency are the primary principles of ecosystems, within which competition between populations stands as a subcategory that serves to maintain this interdependency in a way that sustains the balanced relationship of each population in relation to the whole. (Ibid.)

Clearly, there is nothing in this interpretation of nature that is ultimately disturbing. Competition only *appears* to be competition; it is really cooperation, symbiosis, and interdependence. Natural communities foster mutual aid and eschew hierarchies—there is no "king of the forest," Ruether reminds us, Disney notwithstanding (ibid.). Predation, in the rare instances in which it does occur, might seem unpleasant, but death creates new life and keeps nature balanced and within its proper limits. Given this account of evolution, one wonders why Darwin expressed any qualms whatsoever about publishing his theory.

Humans alone, in disobeying nature's limits and engaging in competition that is not conducive to interdependence, thwart nature's processes and upset its balance. *Our* communities operate on a principle of competition unknown in nature, one of "social hostility and competition for resources" (54). Ruether draws a clear distinction between humans' hostile competitiveness and the competition as maintenance of order that characterizes the animal world. In her efforts to contrast nature's generally beneficent arrangement with the dysfunctional state of human society, she even goes so far as to make the bizarre claim that animals *never* kill humans with the intent of eating them: "Although occasionally a trapped and enraged animal may kill a human, it does not do so for food" (ibid.). Ruether's claim here is simply incorrect, but it serves to underscore the differences between our own practices and the natural ethic we ought to adopt.

THE "SCIENCE" OF GAIA: EARTH AS ORGANISM

Throughout the book—and of course, in the title—Ruether draws upon the concept of Gaia and the Gaia hypothesis, a theory developed by scientists James Lovelock and Lynn Margulis to describe the processes that created and appear to maintain systems of life on earth. The Gaia hypothesis has enjoyed widespread popularity among lay ecologists and environmentalists because of its emphasis on the earth as a single, living organism capable of maintaining its own vital functions in a precarious yet ongoing state of balance. In fact, as Joel Hagen observes, proponents of a homeostatic, superorganism model of ecosystems enthusiastically welcomed the Gaia hypothesis—while others dismissed it as "romantic fiction"—when Lovelock presented his findings in the late 1960s and early seventies. One of Gaia's chief supporters in the natural sciences was ecosystem ecologist Eugene Odum, who was attracted to the theory's emphasis on homeostasis as "a fundamental property characteristic of all biological systems."[7] It is not surprising that Lovelock's Gaia concept should resonate with Odum's ecosystem idea, given that "there was a lot of panglossism in James Lovelock."[8]

For Ruether, and many other environmentalists, the Gaia concept represents a quasi-spiritual understanding of the earth. Ruether's environmental arguments revolve around an understanding of nature as a holistic, living system struggling to maintain the balance that is natural to it in the face of disruption from human activities. However, as in her account of evolution and competition, she borrows selectively from the Gaia hypothesis, emphasizing superorganismal themes of mutuality and interdependence in order to support her particular normative use of "science" for ethics. Though Ruether's account of natural systems includes a somewhat modern emphasis on organisms and their physical/chemical environment (biotic and abiotic interactions), her emphasis on balance, unity, and self-regulation remains antiquated. "Gaia," Ruether writes, "is the word for the Greek Earth Goddess, and it is also a term adopted by a group of planetary biologists, such as Lovelock and Lynn Margulis, to refer to their thesis that the entire planet is a living system, behaving as a unified organism."[9] Ruether finds support in the Gaia hypothesis for an ethic of interdependency, arguing that the theory promotes a "new vision of earth . . . a living organism of complex interdependencies and biofeedback, linking biota and its 'environment' of soil, air, and water" (56). Here again, we see her juxtaposing the Gaia perspective with destructive competition of the sort that humans, and humans alone, employ in and against nature. Only competition as earth maintenance is compatible with the Gaia hypothesis:

> Any absolutization of competition that causes one side to be wiped out means that the other sides of the relation thereby destroy themselves. The falsity of the human cultural concept of "competition" is that it is mutually exclusive. It imagines the other side as an "enemy" to be "annihilated," rather than an essential component of an interrelationship upon which it itself depends. (Ibid.)

Gaian ethics undermines a "dualistic negation of the 'other'" but does not dispense altogether with the category of evil. In other words, Ruether believes that it is important to retain basic distinctions of good and evil when analyzing natural processes. If human ethics ought to consist in a conscious emulation of nature, then these moral categories must have some biological meaning and, indeed, they do, she argues. In nature a life force "becomes 'evil' when it is maximized at the expense of others. In this sense " 'good' lies in limits, a balancing of our own drive for life with the life drives of all others with which we are in community, so that the whole remains in life-sustaining harmony" (256). Humans display a reckless disregard of natural goods and evils, as when "dominant males . . . maximize their own lives, both for leisure and consumption, over against other humans." Nature itself defines such relationships as evil, for Gaia shows us that the "wisdom of nature" lies in limits, where no species' proliferation entails the annihilation of other beings (257).

But does the Gaia hypothesis, as formulated by planetary scientists, support a naturalistic ethic of life-enhancing interdependency over destructive domination and competition? A closer examination of this hypothesis reveals significant differences between Gaia's ethical import for some of the scientists who propose the hypothesis and Ruether's normative appropriation of it. Despite her groundbreaking work in the evolution of cooperation in eukaryotes, Lynn Margulis is highly skeptical that biology (including the Gaia hypothesis) can furnish guidelines for human ethics. She emphatically denies that human modes of annihilating competition are unique in evolutionary history. In drawing this comparison between Ruether and Margulis, my intention is not to show that Margulis has gotten all the ethical implications of Gaia "right" while Ruether has got them "wrong." In fact, there are several points at which I disagree strongly with Margulis's caricature of environmentalism; likewise, her assertion that biological science is ethically neutral is not only questionable in its own right but a statement contradicted within the context of the position she defends. Nevertheless, Margulis's discussion of Gaia illustrates the extent to which Ruether has selectively seized upon some aspects of the theory, namely, those amenable to her ethic. As in her discussion of evolution, Ruether overlooks those features of the Gaia hypothesis that could undermine the natural ethic she believes we should incorporate in human societies.

THE GAIA HYPOTHESIS AND ENVIRONMENTAL ETHICS

Ruether is correct in claiming that Lovelock and Margulis propose that the earth, as a whole, seems to exhibit a system of checks and balances. The earth's own life-forms, they argue, appear to act in such a way as to maintain optimal living conditions on the planet. Our climate offers one compelling example. During the earth's entire history, the sun's output has not remained constant but has increased as much as 30 percent.[10] One would expect that during the first half of our planet's existence, earth would have been cold, barren, and largely unable to support life. Yet this has not been the case. The earth's temperature has remained fairly constant and favorable to many different life-forms throughout its history. Like an individual organism, earth appears able to regulate its "body temperature," despite fluctuations in its own environment. Life on our planet has somehow maintained the conditions favorable to its own existence.

Lovelock at times concedes that Gaia has both scientific and spiritual import.[11] However, Gaia scientists remain skeptical that natural systems can generate normative guidelines for human relationships with nature and the earth as a whole. For one thing, if the earth constantly maintains its very life systems in some sort of equilibrium, why should we worry at all about disrupting nature's systems? It is not always clear that the Gaia hypothesis provides a mandate for changing our current practices. Although it is possible that humans could reduce Gaia's variety of responses to disturbances, Lovelock argues that humans cannot drastically alter the earth: "Man's present activity as a polluter is trivial by comparison [to what other lifeforms have accomplished]" and we "cannot thereby seriously change the present state of Gaia let alone hazard her existence."[12] Thus we might interpret the Gaia concept as an assurance that earth will be able to recover from anything humans do.

Margulis (along with Dorion Sagan) has, in fact, harshly condemned some factions of the environmental movement. She is especially critical of what she sees as the "salvationist rhetoric" embedded in much of the conversation about human responsibilities toward nature: there is something distinctly Christian, "even Puritan," she notes, in our talk about *saving* the planet. Nature is "already out of our hands," she argues; we are not earth's savior.[13] More to the point, her understanding of Gaia contrasts sharply with Ruether's vision of balanced nature and her corresponding claim that natural competition is a subcategory of generally cooperative Gaian interdependence.

GAIA AND ETHICS

Margulis regards environmentalists' conviction that we are destroying nature as both a form of hubris and a kind of "cryptotheism" (348). It is simple arrogance, she argues, to believe that we can either save or destroy the world. Some environmentalists would have us believe that

> the wreaking of such havoc on the environment is unparalleled in earth's long history. This is a kind of negative theology making us, if not God's chosen ones, then his prodigal sons; in any case, as good guys or bad, we remain the stars of the evolutionary show. (349)

But evolutionary reality points to the "deflating fact" that "we have been preceded in our massive ecocide by other life forms" (ibid.). Other organisms have altered the environment by "competing for resources" and "polluting environments" to a greater degree and for a much longer time period than we have. It is simply not true, Margulis states, "that our rapidly multiplying, change-engendering life-form is the only one ever to cause mass extinctions of fellow organisms" (351). Once we realize this, we understand that we cannot "save" the earth (in the sense of protecting all life) unless we were somehow to "stop evolution" altogether (ibid.).

Margulis completely rejects Ruether's dichotomy between nature's "good" competition and humans' "destructive" variety. Our behavior is essentially no different from activities of other species, such as bacteria:

> Like the ecocidal rampage exacted by the violently fast spread of ancient strains of photosynthetic bacteria, our expansion across the surface of the planet has created environmental havoc, and wholesale biological destruction, in our wake. Like those cyanobacteria we have polluted, we have murdered, we have slaughtered with laughter and pride. Like them, we are not good or evil. (359)

We cannot look to nature for ethics that will guide our encounters with other organisms. The turn to nature leads only to an "onset of Dionysian nature madness induced by lack of guidelines" (361). The religious interpretation of nature as needing salvation is merely an attempt to mitigate the shock of "biology's amoral status quo." Some species have been, and will continue to be, completely wiped out in the process of evolutionary competition, she argues, and while these losses may sadden us they are merely the "trimming" that occurs in an "expanding corporation" (358).

The difference between Margulis's metaphor of ruthless, corporate cutbacks and Ruether's harmonious, life-sustaining community is striking. It suggests a radically different understanding of the role of competition and the likelihood that all beings naturally flourish. In Margulis's view the general tendency of all life is to propagate beyond its limits and "in doing so, spoil the environment" (359). This tendency inevitably engages species in a struggle against all other organisms likewise trying to maximize their numbers. If a species can survive and increase its numbers, eliminating other organisms in the process, it has succeeded in Darwinian terms. This is all part of "life's inescapably murderous legacy" (350). Ruether presents a vastly different scenario. She believes that the *general tendency* of nature as a whole is to maintain limits. Organisms—other than humans—remain within their proper niches and do not overrun or dominate others. Whereas Margulis sees struggle and death as directly entailed in the propensity of all life to maximize its numbers, Ruether holds that the Gaian principle of interdependence works to support life, averting widespread death and disaster by creating interactions that are "mutually life-enhancing for both sides."[14] For Ruether, the Gaia hypothesis proves the "falsity" of competition "that causes one side to be wiped out."[15] In other words, Margulis and Ruether disagree fundamentally about the goal toward which life as a whole "aims." Yet, remarkably, both present their accounts as consistent with the Gaia hypothesis.

The Gaian view of life, according to Margulis, provides no background environmental stasis against which to compare our "disturbance" of nature.

In this sense her interpretation of the Gaia hypothesis puts it closer to chaos, disequilibrium theory than to Romantic, organismal perspectives that Ruether embraces. In fact, Margulis cites data from "chaos mathematics, disequilibrium thermodynamics, and complexity studies" in order to discredit environmentalists who mistakenly assume that the planet has been thrown out of its normal state of balance.[16] Perhaps nature is approaching a turning point marked by greater complexity, rather than an end, she suggests. In any case, life in *some* form will continue with or without us, Margulis argues, and that fact is neither good nor bad from the perspective of Gaia; it is simply another stage in her evolution. Once we realize this we can appreciate her claim that environmental conservation on an evolving planet "is ultimately a lost cause."[17]

Of course, Margulis has not necessarily arrived at the only "correct" ethical interpretation of the Gaia hypothesis. Her claim that biological data are morally neutral is as questionable as Ruether's assertion that nature operates according to an ethical principle of interdependency that humans should adopt. Both arguments, I think, overstate the case.[18] But Margulis's picture of nature is the more complete one: she agrees that the earth behaves holistically, as a global ecosystem whose properties appear in some sense organismal—"more akin to systems of physiology than those of physics."[19] But this vision of earth does not translate into a denial of Darwinian struggle and competition. In Margulis's view, earth is neither a dead machine nor a solicitous mother. The Gaia hypothesis is merely an instance of the theory of natural selection occurring on a global scale; "Gaia is Darwin's natural selector," and some species simply do not make the cut.[20]

Recall Ruether's argument that nature involves three levels of ethical interrelationships: those between humans, those between humans and nature, and those presumed to exist among nonhuman life-forms in nature. Her contention is that the same ethic ought to exist at all levels. Perhaps more accurately, she believes that clues to the proper ethical relationships *already exist* among nature's creatures. Simply put, humans can learn to live *with* nature by learning to live *as* nature lives. In evaluating her argument, then, we first have to consider whether she is correct in claiming that proper ethical relationships actually exist in nature. Second, we have to ask, given the sort of relationships that actually exist between organisms in nature, whether it makes sense for humans to adopt them as normative.

Ruether would answer in the affirmative to both these questions, claiming that "both nonhuman nature and also humans have an inherent capacity for biophilic mutuality."[21] However, as we have seen, she has selected particular features from the available stock of "natural" evidence that has bearing on her claims. To widen her scope of empirical data to include a more accurate—and more Darwinian—picture of nature would jeopardize the ethic of loving and just relationships that she wishes to promote.

But, for the moment, let us assume that Ruether is correct in her normative reading of nature. The implication for the human-nature relationship is that humans ought to regard nature and its inhabitants in the same spirit of mutuality and cooperation. In fact, however, Ruether urges *more* than this for humans' treatment of nature—she issues a plea for "healing" the earth. But why should humans heal nature if nature already functions as a life-enhancing and sustaining "healthy web of life" (1)? Isn't nature already, in a sense, healed?

Presumably what Ruether intends is that humans should heal the damage that we, specifically, have done to nature—anthropogenic damage that can clearly be traced to our disregard and misuse of nature, such as "proliferating population . . . soil erosion, deforestation, the extinction of species, air and water pollution" (259). The role of humans as earth healers does not place us outside of nature, she argues. Rather, healing is itself an expression of our acknowledgment that we are inextricably connected to nature. Put differently, she denies that reenvisioning our role as earth healers aligns us more with God than with nature. The idea that humans and humans alone image God—an attitude that has led to dominating behavior—must be purged from our consciousness, she argues, if we are to address environmental damage. Nothing short of *metanoia*—a change in our consciousness—is required in order to correct our wrong relationship with nature.

In support of humans as earth healers, Ruether draws on a biblical "covenantal vision" that complements the ecological ethic emerging from science. (As we will see, Michael Northcott makes similar claims regarding biblical covenant concepts and ecological ethics.) The covenantal strand in Christianity illustrates that the Bible is not as "antinature" as it is often portrayed to be (207). Hebrew thought understood the covenantal gift of land as something that cannot be held apart from relation to God (211). Specific ethical guidelines for the treatment of animals and land accompanied this gift. Weekly cycles guaranteed days of rest for all living things; the seven-year and Jubilee cycles also established periods of recovery for land and animals. These cycles were designated as times of "periodic righting of unjust relations" (213). The covenantal prescriptions for treatment of nature affirmed "that humans and other life forms are part of one family, sisters and brothers in one community of interdependence." Like the ecological ethic, this metaphor understands humans as an integral part of a larger natural community while also supporting the view that we must act as earth's caretakers.

Yet the *covenantal* concept seems to lead to much stronger claims regarding human obligations toward nonhuman life than does the ecological community account that Ruether derives from nature. As we have seen, understanding our obligations to nature in terms of covenant places upon us a burden of maintaining "right relations" with our fellow creatures. Jubilee laws are significant as gestures toward redemptive eco-justice, Ruether argues, but they are temporary solutions. Obeying covenantal guidelines will not bring about a permanent eradication of wrong relations, for evil will always creep back into our dealings with nature because of human sin and arrogance. A "once-and-for-all" restoration of right relations awaits a "final fulfilment of the covenant of creation" bringing "peace between people and healing nature's enmity" (ibid.). In other words, the ultimate significance of the covenant lies in its messianic fulfillment in the Peaceable Kingdom foretold in Isaiah, when "even the carnivorous conflict between animals will be overcome" (ibid.). In the meantime, humans are accountable to God to "care for and protect the vast panoply of life forms" on earth (222). But it is precisely here that the covenantal metaphor for human responsibility appears at odds with the ecological community ethic; Ruether's own message of ecological interdependence and mutuality seems to have gotten lost. According to the covenant concept, human responsibilities toward nature *do not* consist merely in emulating ecological processes and understanding ourselves as one inextricable strand in the larger web of life. The covenant role implies that we are more than just members of the ecological community, more than mere "latecomers to the earth, a very recent product of evolutionary life" (5). Rather, we are now "guardians" of nature, responsible for restoring it until the time of the eschaton.[22] Ruether's turn to the vision of a peaceable kingdom that will ultimately remove all "enmity" and "conflict" is particularly odd, considering that her ecological model denies enmity even exists in natural communities. But, more significantly, why should humans assume the position of guarding and restoring nature when it is nature that presents us with the proper ethical norms? Ruether seems unable to decide how "good" nature is and how "evil" humans are. Moreover, although she argues elsewhere in *Gaia and God* that it is a mistake to equate sin with changeablity and death (in nature and in human life), and that such perspectives have led to an "earth-fleeing ethic and spirituality," her invocation of the peaceable kingdom suggests that natural processes are, in and of themselves, somehow evil or in need of correction.[23] Whether or not Ruether means to suggest that humans should "heal" the conflict between predator and prey,[24] her association of covenantal duties with meting out eco-justice is problematic in light of her ecological ethic. This is not to say that Ruether must formulate an ethic that is purely scientific or ecological in its origins or that religious metaphors have no place in an ethic that is informed by biological data. But she needs to establish greater congruence between the ethical implications she derives from nature and the biblical themes she develops in support of human responsibilities toward life. Taking her own description of interdependence more seriously would imply a more modest role for humans than that of arbiters of justice and purveyors of peace and health in nature.

Ruether is not alone in desiring to bring together an ecological ethic and a Christian model of care for others. Sallie McFague shares her conviction that the Christian perspective—or aspects of it—is highly compatible with a scientifically informed ecological ethic of interdependence. As we will see, McFague's ecological theology suffers from many of the same shortcomings, both in terms of scientific accuracy and in terms of genuine applicability as an environmental ethic.

MCFAGUE'S ACCOUNT OF CHRISTIAN LOVE TOWARD NATURE

McFague's arguments assume a different starting point from Ruether's, but many of her claims are similar. She argues that a Christian ethic of love and care—amplified in certain ways by inferences from ecological science—can consistently be extended to nonhuman nature. Her argument involves a dialectic between ecological norm and Christian message, showing the affinities between the two and noting how each perspective gains support from the other.

We will begin with a fairly detailed discussion of McFague's understanding of metaphorical theology. It is important to place her arguments in *Super, Natural Christians*—her work most explicitly devoted to human-nature ethics—within the context of her larger project. As we will see, McFague's environmental ethic lacks a solid scientific grounding and, like Ruether, she appears much more comfortable with "ecological" themes of benign interdependence than with evolutionary concepts. Moreover, I argue that based on McFague's definition of metaphorical theology, her argument fails to meet certain criteria for what constitutes an adequate choice of metaphors. That is, according to her *own* standards of what a well-crafted metaphorical theology should accomplish, her arguments fall short.

METAPHORICAL THEOLOGY

In *Super, Natural Christians* McFague further develops lines of argumentation seen in her earlier works such as *The Body of God* and *Models of God* (both of which place at least a nominal emphasis on the "ecological" perspective). Much of her work adopts what she calls a "heuristic perspective" or "experimental theology," and *Super, Natural, Christians* is no exception. McFague argues that living in an ecological, nuclear age raises unique and difficult problems for human life. The possibilities for our own destruction, and the necessity of recognizing our dependence on a fragile natural world, are brought into sharper focus. Coming to terms with life in an ecological, nuclear age calls for the development of a new consciousness, a new way of conceiving of humans, nature, and God. This is what she proposes to offer.

Traditional ways of conceiving our relationship to God and nature are obsolete, even dangerous, in the modern world, McFague argues. She shares Ruether's concern with human alienation, but she is especially interested in the ways that our *language* about God and nature have alienated us from both. New ways of mythologizing God and the world can challenge dualistic, hierarchical, and anthropocentric attitudes that have characterized Christianity for centuries. We can reconceptualize our relationship with God by shedding old metaphors that depict God as male, monarchical, and transcendent, for example. McFague suggests that we think of God, instead, along the lines of a mother, a lover, or a friend and imagine the earth as the body of God.[25] By reenvisioning our relationship with God, we can begin to alter our relationships with others and with the planet as a whole. Models of God as mother, lover, and friend "imply views of the creative, salvific, and sustaining activities of God that are radically different from what we find in the monarchical model." New metaphors increase our awareness "of the interdependent network of life, with God at its center as well as at every periphery."[26] Experimenting with new metaphors for God and nature has great heuristic value—we can "try on" different metaphors and see where they lead in terms of the kind of thinking and action they generate.

All metaphors, she argues, have a "no sense" or nonsense quality to them: metaphors simultaneously depict things as they are and are not. This nonsense aspect of metaphor gives us room for play. Given that all metaphors are, in a sense, inaccurate, we should seek to adopt those that confer certain practical advantages and encourage better relationships. "A theologian's job," McFague argues, "is to help Christians think about God, other people—and nature—so that we can, will, act differently toward them."[27]

Yet McFague's argument about God and nature is not intended to remain on a simply experimental, metaphorical, and heuristic level. Despite the nonsensical and "playful" nature of metaphorical theology, metaphors are bound to certain criteria: they should be adopted in light of empirical considerations and evaluated in terms of the appropriateness of the ethical behavior they encourage. Thus, in keeping with the first criterion, McFague seeks to reform Christian theology in a way that is "informed by and commensurate with contemporary scientific accounts of what nature is."[28] The articulation of new metaphors is aided by sources outside of religious traditions, including "the natural, physical, and social sciences."[29] Metaphorical theology must stand up to at least *some* degree of scientific scrutiny, even though metaphors are by definition not accurate descriptions of what they image. Second, a good metaphor also generates ethical prescriptions that are appropriate for the context in which they arise. The whole point of play with metaphors is to find one that, in a sense, works the *best* in a given context. All metaphors are nonsense, she argues, but some are better than others. Herein lies McFague's pragmatic justification for selection of metaphors. "Truth," she contends, "is what is effective in a nuclear, ecological age." A good metaphor is one that is good for us and nature as a whole.[30]

A third criterion for our choice of metaphors also emerges regarding the use of metaphors in the Christian tradition. We ought to give preference to metaphors and models that, while having a kind of shock value to them, still maintain some degree of continuity with the traditional symbols and interpretations they replace. Thus, for example, in proposing that we envision nature as God's body, McFague points out that in Christianity there already exists a significant amount of "body language"—biblical stories that underscore the importance of bodily life and health, for instance. The idea of the earth as God's body is both novel and consistent in the context of Christian thinking. The novelty of the metaphor lies in its revelatory implication that the earth, as God's body, is vulnerable and at risk. This insight radically changes our traditional understanding of God while still maintaining important links to traditional Christian ideas of embodiment. With these criteria for choice of metaphor in mind, we arrive at the following approach: metaphorical theology should offer models of God and nature that constitute both a break from and a link with tradition; metaphors should be scientifically informed, and they ought to generate desired ethical values, the truth of which lies in their effectiveness in a modern context. "Theology," she reminds readers in her recent work, *Life Abundant*, "is a functional activity" the goals of which are "practical and pragmatic."[31]

This understanding of metaphorical theology forms the basis of McFague's "ecological ethic" in *Super, Natural Christians*. Her thesis is that Christians need to reenvision nature and its inhabitants as "subjects" in a subject-subjects relationship. We should treat all others, human and nonhuman alike, as "Thous" rather than objectifying them as "its." McFague's exhortation to regard all beings as Thous is implied already in her earlier works, which, likewise, seek to overturn hierarchical and dualistic relationships, but *Super, Natural Christians* addresses the human-*nature* relationship more explicitly and in greater detail than her previous writings. Given that our way of understanding nature is a cultural convention, the good news is that we can change those conventions. "We think in terms of major metaphors and models that implicitly structure our most basic understandings of self, world, and God."[32] Nature, she notes, has often been the "neglected third" in the God-human-nature relationship, and thus she sets out to explain to Christians exactly how they are to love nature.[33] In *Life Abundant*, she puts the point more forcefully, suggesting that a loving, ecological theology is essentially the same thing as liberation theology. Here she articulates

> an "ecological theology"—one that rests on the relative absolute of giving glory to God by loving the world; that understands its context to be the well-being of the planet and its subjects all creatures; . . . one for the entire cosmos with all its creatures, human and otherwise. An ecological theology of this sort could be a liberation theology for twenty-first-century North American Christians.[34]

But first, we have to rethink some biblical stories and Christian symbols, she argues, for while an "ecological liberation theology" contains "roots of both the Hebrew and Christian traditions" it is also novel because until recently these traditions have focused too exclusively on humans alone.[35]

Fortunately, there are already elements of the Christian tradition that we can build upon. Christians have a model of relating to one another and to God that can "simply be extend[ed]" to nature. We should "relate to the entities in nature *in the same way* that we are supposed to relate to God and other people—as ends, not means, as subjects valuable in themselves, for themselves."[36] What makes this Christian model even more compelling (and fulfills one of her criteria) is that such an approach is consistent with modern "ecological science." The two "models"—the Christian and the ecological—point toward the same ethical message: "The ecological model says that the self only exists in radical interrelationship and interdependence with others and that *all* living and nonliving entities exist somewhere on this continuum."[37] The Christian model values all others as subjects, ends in themselves, and understands the individual as an inextricable part of a community of care.

In the discussion that follows we will examine this Christian model and the ecological perspective in greater detail, paying particular attention to McFague's claim that the two are similar and mutually reinforcing. We will then turn to a critical, and scientific, evaluation of McFague's description of the ecological model. Like Ruether's ecotheology, this account of nature bears little or no resemblance to the modern evolutionary perspective she claims to adopt. McFague simply highlights certain "ecological" dimensions of nature that seem to fit the Christian ethic of care and love she has in mind. I will also attempt to apply the subject-subjects approach to nature (an approach that, according to McFague, finds equal support in the Christian and the ecological perspective). We will test this approach with regard to both cultivated and wild nature, in order to see whether it can succeed as an environmental ethic. My contention is that McFague's account of loving nature fails at least two of her own criteria for experimental theology: it is not adequately informed by science and it does not fulfill the requirements as a pragmatic definition of truth.[38]

THE CHRISTIAN MODEL

If, as McFague asserts, metaphors and models shape our basic assumptions about nature, we must first excavate the models embedded in Christian theology before we can seek to reconstruct them. The basic model of nature that has often pervaded Christian thinking—and the one that, historically, has been encouraged by scientific, mechanistic interpretations of the world—has been the subject-object model. This perspective regards others as things, as mere means to the ends of the knowing subject. The subject-object model is "dualistic, hierarchical, individualistic, and utilitarian," and has been harmful in its treatment of nature as a dead machine with no intrinsic value.[39] McFague therefore proposes that we try a different model, one that "sees everything, *all others*, as subjects" (9). This fundamental, radical shift in perspective is necessary before we can hope to halt the destruction of nature.

Here the Christian tradition, as well as modern science, can help. An alternative perspective on nature already exists within the Christian tradition (especially the ministry of Jesus) and, increasingly, within emerging scientific literature about the natural world and our place in it. Within the Christian tradition the liberation paradigm, rooted in Jesus's ministry, offers us a "strong candidate" for addressing environmental problems. Liberation thinking understands that human and natural salvation are one and the same. The new liberation-oriented Christian ethic that McFague proposes would entail simply extending "the destabilizing radical love we see in Jesus' parables, healing stories, and eating practices *to nature*" (14). The ministry of Jesus underscores the love of Christians for all beings, but especially those who are most needy, vulnerable, and oppressed. Nature, McFague argues, should be understood as the "new poor . . . the earth others who are suffering bodily oppression" (6). This shift in perspective exhibits both affinities with, and significant departures from, the inherited Christian tradition.

McFague defines three important steps in developing a Christian nature praxis. The first is a "deconstructive" one: Jesus's parables instruct Christians to overturn hierarchies of rich and poor, powerful and weak. Once these hierarchies are dismantled, the next step is one of "reconstruction," drawing on the parables of Jesus' healing. Such healing acts, extended to nature, "underscore that salvation means the health of bodies, that the deterioration and dysfunction of the ecosystems of our earth are affronts to God, who desires the well-being not just of humans but of all creation" (15). Healing the sick is central to Jesus's ministry, McFague argues, and reflects the importance of physical health as "a basic right" (16). (At this step in her argument, McFague already alludes to the concept of ecosystem health. We will look at her understanding of "ecology" in a moment.)

The third step in Christian praxis is a "prospective" one, highlighted in the eating practices of Jesus. The parables reveal that Jesus ate with societal outcasts, an act that symbolizes his solidarity with *all* others—including sinners, tax collectors, and prostitutes. Again, these stories point to the importance of bodily health and the demand for basic justice for all beings: "all bodies," McFague concludes, "need to be fed" (15). Furthermore, Jesus's eating practices are also a symbol of hope, orienting Christians toward "the eschatological banquet when all creation will be satisfied and made whole" (ibid.). Narratives regarding Jesus's eating practices illustrate that he cared for *all* others—even those rejected by society—and that his care for them involved their bodily as much as spiritual well-being. Our relationship with nature should be modeled on Jesus's praxis as illustrated in these three steps. Viewing *all* life as subjects and not objects is consistent with the ideals of Christian love embodied in these stories and engages us in the struggle against the oppression of nature.

MCFAGUE'S ECOLOGICAL/EVOLUTIONARY PERSPECTIVE

The subject-subjects model also resonates with what McFague calls the "ecological, evolutionary perspective" or (more commonly) the "ecological model" (49–53). She begins her discussion of nature by acknowledging that the term *nature*—like *Christian* and *spirituality*—connotes many things, not all completely positive. We "have a choice about how we are to interpret nature," she warns, "and the choice we make will be important" (20).

How, then, does McFague interpret nature? Essentially, she argues, "nature is everything." Nature includes "human beings, but also black holes, electromagnetic fields, dirt, DNA, gravity, death, time and space, apples, quarks, trees, birth, microbes, tigers and mountains" (16). Since this definition of nature can be overwhelming, it is helpful to understand nature in two basic senses—a larger sense (the "big picture") and the smaller, more localized sense. The small picture is nature in its particularity—the places with which one is familiar, the landscape (perhaps simply the backyard) of one's childhood. The big picture of nature is "the picture of the physical world coming to us from postmodern science and ecology," one that is "characterized by evolutionary change and novelty, structure as well as openness (law and chance), relationality and interdependence, with individual entities existing only within systems, systems that can be expressed by the models of organism and community" (20). So far, so good. McFague has indicated that she will give at least some consideration to positive and negative aspects of nature, to features of evolution (chance, change) as well as to ecological themes of organismal community and interdependence. The evolutionary, ecological model, she argues, should play a role both implicitly and substantially in Christian thought. The big picture of nature is the worldview that underlies the smaller picture—i.e., all our daily encounters and experiences of nature. This worldview must be accepted, must be taken seriously, she argues, because it represents "*the contemporary picture of reality*" (21). Moreover, the ecological, evolutionary understanding is one that Christians should fully incorporate into their thinking about nature, not only because it is, in some sense, reality but because "this view of reality is highly compatible with the spirituality of Jesus's ministry, for it sets us in a world of radical relationality at all levels . . . of our existence" (22). It is important that Christians see the world as it really is, McFague argues, because we cannot love what we do not know: "A certain kind of knowing is how Christians ought to love nature" (32).

But what does it mean to "know" nature? Is it enough to be able to identify the plants and birds in one's backyard? Knowing a particular natural place in some detail is crucial. But one must also have a grasp of the larger scientific picture of our planet. Both perspectives are necessary, but they are still not sufficient, McFague argues. How one understands nature depends on how, exactly, one looks at it. McFague distinguishes what she calls the loving eye from the arrogant eye. The arrogant eye sees all things only in relation to the self, objectifying all it sees as potential acquisitions, while the loving eye values complexity, mystery, and difference in all things and beings. The loving eye, in short, acknowledges and accepts others as subjects—"and this model is extended to the natural world and its lifeforms" (34). Love sees all beings as they really are. Christian love, then, embraces reality: love is "simply facing facts" (36). The subject-subjects model involves looking with the loving eye, and this model "fits the data better" than other models in that it corresponds to "the postmodern view of reality" (96).

What are the data, the "facts," that support this perspective and how did previous views of reality differ from our current one? McFague offers an historical sketch of models and paradigms regarding nature. The modern approach, which she dubs the "ecological model" has affinities with past models but also constitutes a break with them in important ways. The modern ecological model is essentially the *same* model as the subject-subjects approach. Both point to a community ethic of interdependence. Let us look at McFague's account of the "recent" emergence of the ecological model.

PARADIGM SHIFTS

McFague identifies three important shifts in our thinking about the natural world, namely, 1. the medieval paradigm, which in its implications for subjecthood in nature resembles the modern view in some ways, 2. the Enlightenment shift in perspective, which resulted in a subsequent loss of subjecthood for nature (seeing nature as mechanized rather than living), and 3. the modern/ecological return to subjecthood in nature, which is now replacing the mechanical view.

The medieval view, like the modern ecological model, understood nature as fundamentally ordered around principles of relationality and interconnectedness. Humans, McFague argues, once experienced a profound closeness with nature and found within it one means of being closer to God. Symbolic, religious meaning pervaded nature. But unlike the modern ecological view, the medieval view held that the human occupied the "privileged place" in a thoroughly hierarchical arrangement with God at the top, and animals—created for human purposes—occupying the lower rungs (ibid.). It was a world characterized by "harmony, proportion, and stability," but also an inherently hierarchical one, a "vertical grid of decreasing value" (49–51).

Modern ecological interdependence echoes the medieval model, McFague argues, but does so in ways far more "in keeping with the contemporary scientific picture of reality" (51). The medieval model did not respect the intrinsic value of all things but, rather, perceived them anthropocentrically, as lesser beings ordered toward the benefit of humans. As such, the medieval model tended to eclipse the value of individual beings, seeing the larger whole as the important unit. McFague argues that the modern ecological view maintains the *individuality* and *distinctiveness of the self* even while seeing the self as constituted by its relationships, by its radical relationality to all other beings. The modern ecological view also differs from the medieval one, she stresses, in that its understanding of subjecthood is nonhierarchical. The ecological view is more radical than the medieval view in its emphasis upon individual beings as possessing subjecthood. Note that, at the same time, the modern ecological perspective also affirms the value of the totality of nature—the entire planet—as a single living organism, as the medieval view had recognized long ago.

The intervening Enlightenment model of nature as an efficient machine stripped the natural world of its subjecthood. Enlightenment sensibilities popularized the view that humans are "unrelated to nature"—an idea that would never have occurred to a medieval peasant and one that strikes us, with our modern ecological perspective, as completely absurd (48). This notion brought a "profound change in sensibility, manifested in seeing the earth in mechanistic rather than organic terms, in the rise of science and technology, in the Enlightenment's confidence in human rationality."[40] Destructive patterns of thought established by the Enlightenment/mechanical model of nature have had far-reaching consequences and continue to be perpetuated by both scientific and religious worldviews, McFague argues. The Enlightenment model marked the advent of subject-*objects* thinking and its legacy haunts us—and our natural environment—to this day. Only recently, with the ecological model, have we once again come to understand nature in pre-Enlightenment terms.

When we consider McFague's overview of the paradigm shift from medieval views of nature, to mechanistic models, to the current ecological model, it is astonishing to realize that she never discusses another important paradigm shift: that provided by Darwinian evolution. Only one mention of evolution appears in this historical overview, when she reiterates that the present ecological view supposedly incorporates evolutionary theory within it: "Ecological interdependence, built upon evolutionary theory, takes full account of the negativities of existence; in fact some see it as nothing but an advertisement for the survival of the fittest and the accidental, arbitrary character of existence."[41] Like Ruether, McFague subsumes evolution under a larger principle of ecological "interdependence;" it plays at best a supporting role in the truer narrative of community. Also, like Ruether, McFague instantly associates evolution with certain "negativities," primarily the "survival of the fittest" mentality. She then quickly asserts that this mentality is not justified, due to the greater body of evidence in nature for "a view of kinship, interdependence, and relationality so radical that we can speak, metaphorically, of the earth as the common mother of all that exists on our planet."[42] The "negativities" of natural selection are never explored, except for a brief reminder that they are not so significant as to justify a *generally* pessimistic outlook. This disclaimer is given along with an assurance that these negativities (whatever they consist in) have already been incorporated into the ecological perspective.

If kinship and radical relationship with nature are the key components in the paradigm shift that ultimately overturned the Enlightenment view, then why not stress the importance of the evolutionary paradigm? It was, after all, Darwin's theory of evolution—not medieval hierarchies or Romantic reflections on interdependent communities—that established the true kinship between humans and animals. Moreover, Darwin's works dealing with animal intelligence and emotions directly challenged many of the "mechanistic" Cartesian assumptions of the science of his day. Surely if ever there were a dramatic shift in our understanding of our place in nature, Darwin had something to do with it.

The ecological model in which McFague identifies traces of the medieval view is, in fact, closer to the Romantic ecology of the eighteenth and early nineteenth centuries (as well as ecosystem concepts of fifty years ago) than it is to a cutting-edge, postmodern scientific perspective she professes to adopt.[43] Romantic accounts of ecology were themselves a conscious reappropriation of pre-Enlightenment views. As we have seen, the pre-Darwinian ecologists sought to reanimate the world, to reclaim it from Enlightenment mechanization of nature. Before Darwin many naturalists approached the scientific study of nature "armed with a much older, basic intuition that nature was alive." Central to this view

> was what later generations would come to call an ecological perspective: that is, a search for holistic or integrated perception, an emphasis on interdependence and relatedness in nature, and an intense desire to restore man to a place of intimate intercourse with the vast organism that constitutes the earth.[44]

The point is that the "new" paradigm McFague describes is in fact not so new at all. Her reluctance to feature the evolutionary paradigm in her historical overview is significant. McFague's association (and Ruether's, for that matter) of evolution with the phrase "survival of the fittest" may partially account for her avoidance of Darwinism. A survival-of-the-fittest understanding of evolution would seriously undermine the ethic of preference for the "needy" and "oppressed" in nature that McFague wishes to promote. If a Christian love ethic extended to nature gives preference to whose who are oppressed—those who are suffering, starving, wounded—a survival of the fittest account of Darwinism seems to give preference to the oppressors—the dominant, strong, exploiting class. McFague's ethic, rooted in the "deconstructive" dimension of Jesus's ministry, explicitly seeks to overturn the oppression of the weak and vulnerable by the strong and powerful, whereas the survival-of-the-fittest "ethic" legitimizes and naturalizes that oppression.

Despite McFague's plea to Christians to take seriously the evolutionary, ecological perspective, her account of nature, and the ethic that emerges from it, suffer greatly from an avoidance or miscontrual of basic "facts" about the world. She begins with a nominally "ecological, evolutionary" perspective, but increasingly the word *evolution* begins to fade from her discussion. She introduces the term *evolutionary* in the first few chapters but henceforth speaks only of the "ecological model" or the "ecological perspective." We find an emphasis, once again, on the theme of interdependence of nature and a corresponding "ecological ethic" of community and cooperation. The scientific "reality" that McFague would have Christians confront is a one-sided version. Furthermore, as we will see, her attempt to describe scientific reality in terms that render it consistent with Christian praxis results in an inappropriate ethic toward nature that might ultimately create more problems than it would solve. Before taking a closer look at these criticisms, it is important to get one caveat out of the way regarding McFague's metaphorical theology and her use of models.

ON MODELS, METAPHORS, AND REALITY

At this point I suspect that McFague, and some of her readers, might object that I have misunderstood the point of a metaphorical theology. She is, after all, simply proposing a new way for Christians to *conceptualize* nature. Her project is intended to be heuristic, analogical, and experimental; it performs a pedagogical function. In fact, in the opening pages of *Super, Natural Christians* she issues what appears to be a disclaimer along these very lines:

> This book, then, is meant to be a laboratory to experiment with, try on, a new sensibility, one that sees everything, *all others*, as subjects. As much as a book of this sort can do, it is meant to be an experience, a test case, of whether a proposal for seeing everything through the lens of subjecthood is helpful and perhaps even necessary for Christians in our time.[45]

Given that we always think in terms of models (which are by nature "relative and inadequate"), McFague asks, "Is it not better (not perfect) to think of the rest of nature more or less *like ourselves*, rather than utterly unlike us?"[46] She is simply reorienting us, not claiming to offer a description that corresponds precisely and in every detail to natural, scientific reality. Yet, even granting that McFague's account is offered as a test case, this experiment is on the whole unhelpful, as we will see. The argument that her project should be evaluated in a spirit of experimentation is, in any case, not a sufficient defense. As Northcott observes, McFague's entire approach to theology and ethics begs a number of questions that cannot be dismissed simply by pointing out that religious language is metaphorical and analogical. "Environmental philosophers do not resort to this linguistic sleight of hand," he argues. "Why then should ecotheologians?"[47]

McFague's theology, while metaphorically driven, does not simply deal in metaphors. Her frequent and emphatic reference to the importance of knowing and understanding details and "facts" about nature, her professed desire that Christian theology incorporate the "reality" of current science, and her definition of truth as "effectiveness" all point toward an intention to present a Christian environmental ethic that has real, practical—not merely theoretical or aesthetic—value. "Science," McFague writes, "helps us to become travelers to the new world of nature, to replace our ignorance and half-truths with genuine and accurate information."[48] Moreover, she claims that the model of human-nature relationships that she proposes "results in more humane and healthy practice . . . mak[ing] for immense changes at both personal and public levels."[49] My critique of McFague assumes that such statements about the importance of attaining scientific accuracy and changing environmental practice are intended to be taken seriously.

But what does McFague mean by the term *models*? How is a model different from *reality* (a terms she also uses regularly)? Her use of the "ecological model" is at times perplexing because she speaks simultaneously of this model as one we derive from nature and one we *extend to nature*. Isn't the ecological model "in" nature in some sense? If not, why is it called the *ecological* model?[50] If it doesn't derive from nature, where does it come from?

Most of her discussion of this model centers on its implications for seeing subjecthood in nature; unfortunately, there is less discussion of what the "facts" of nature are that support the "model." Presumably, the scientific reality that undergirds our daily encounters with nature forms the basis of the ecological model. Yet, she argues, this picture of reality is just our "current picture." McFague writes that the "evolutionary, ecological, relational, community model of nature *is the contemporary picture of reality*. It *is* a picture," she adds, "(and not reality 'as such'), but so are the medieval and Newtonian views."[51] Later she claims that the ecological/subjects-subjects model is "a relational model derived from the evolutionary, ecological picture of reality" (38), and again that the ecological model "is *derived* from the workings of natural systems" (107). There is some confusion here. McFague presents the model *itself* as the contemporary scientific picture of reality. Yet, she argues, the model is also *derived from* a picture of reality. What is the difference between the model and the scientific reality upon which it is based (particularly when both are referred to as evolutionary and ecological)? Why bother distinguishing reality from the "models" that correspond to it if both are really, as she maintains, "just models"? Furthermore, what distinguishes "reality" or "pictures" of reality from "reality *as such*"? Does she believe that there is an ultimate, objective reality beyond our "picture" of it?

Presumably, what McFague means to say is that postmodern perspectives illustrate the untenability of speaking of reality as something that exists apart from our constructions of it. All reality is a "version" of reality. But if reality is itself a model or version, then how is it possible to check our models for compatibility with scientific reality, as McFague urges? In fact, I think that McFague uses the terms *model* and *reality* to mean different things in different contexts and that these different meanings are not well defined in her arguments. Her understanding of reality *as* model permits her an escape from criticisms on empirical grounds, because she can always object that she is, after all, simply working with models, not accurate *descriptions* of nature. When told, for example, that the scientific account that informs her ecological model is incomplete or inaccurate, she can respond that she is not making claims about reality "as such": "Remember," she adds parenthetically, "we are always working from within relative, inadequate models, none of which will be perfect" (157–58).

Models and the reality they image are presented as equally provisional in this account. It is not clear why we should adopt the ecological model over the Newtonian, subjects-objects model of nature, unless the former is more "accurate" (a term she sometimes uses), in addition to being a "better" one ethically. This in fact is what she argues: the ecological, subject-subjects model "fits the data" better (96). It is the model "closest to reality" (100). It is not clear how one would judge this without assuming that reality is in some sense "real." Perhaps she means to say that the ecological model fits the "fact" of interdependence.[52] But her interpretation of interdependence filters out the evolutionary dimensions of interdependence, placing exclusive emphasis on mutuality and community. Or, put differently, she fails to separate interdependence as "data" from interdependence as model or metaphor. Interdependence is presented as necessarily and self-evidently good.

McFague's discussion of models and reality obscures the issue of what constitutes scientific accuracy regarding nature. She is right, of course, to say that our picture of reality changes over time, that scientific accuracy is not something attained once and for all: we no longer think of the earth as the center of the universe, for instance, and most of us believe the planet to be more than a few thousand years old. But by placing what she calls "reality" on the same ontological plane as the "subject-subjects model," she blurs the distinction between scientific data, on the one hand, and an environmental ethic that might be generated from it, on the other. (There is also something strange about McFague's repeated demand that we should accept a model of reality that already *is* the current model of reality, as when she urges Christians to make this picture of reality more "commonplace, even conventional" in their thinking about nature. Presumably "reality," given McFague's own account of it as socially constructed, must be something that, by definition, a large number of people already accept!)

The confusion in McFague's account of models and reality may be exacerbated by the fact that her metaphorical theology initially dealt with models of *God*, before her more recent foray into environmental ethics. In an earlier essay,[53] she argues that "no metaphor or model refers properly to God. . . . All are inappropriate, partial and inadequate."[54] Since *all* models and metaphors for God are "improper," multiple, contradictory models do not present a problem in theology as they would if we were seeking accurate descriptions. Metaphors are appropriate in theology, and particularly with reference to God, because "what a metaphor expresses cannot be said directly . . . for if it could, one would have said it directly."[55] Metaphors are an "attempt to speak about what we do not know in terms of what we do know," and thus they are appropriate "*in certain matters where there can be no direct description*."[56] Such matters arise in theology which "is always dealing with improper language, language that refers only through a detour" of description and thus "can always miss the mark."[57] But natural environments, unlike God, are not something of which we have no direct experience or for which we can provide no direct description whatsoever. To be sure, all knowledge, including scientific knowledge, involves modeling and other imaginative attempts at description. Darwin's work, for instance, is full of metaphors and literary devices for describing natural selection. But describing nature is not like describing God, where metaphor is *all* we have to go on, where "missing the mark" is always expected.[58] When McFague directs her metaphorical approach specifically to nature and animals, she encounters a problem of "reality,"—of "data," "facts," and "accuracy"—that did not emerge with the same force in her attempts to model God.

LOVING NATURE ACCORDING TO THE ECOLOGICAL MODEL

As we have seen, McFague maintains that a basic affinity exists between the Christian understanding of love, exemplified by the praxis of Jesus, and the model of the world that has emerged from the natural sciences. She presents the basic features of the "evolutionary, ecological perspective" and proceeds to illustrate what it would really mean "if we were to live within the ecological model."[59] Like Ruether's, this model is mutualistic and symbiotic: "In the evolutionary, mutualistic model, all entities are united symbiotically and internally in levels of interdependence but are also separated as centers of action and response, each valuable in its own beingness."[60] This model suggests that "we human beings and everything else are knit together in intricate, deep webs of interrelationship and interdependence."[61] Ecological relationship is "not dyadic but multiple; it is not oppositional, but interactive."[62]

While holistic, this model is also deeply concerned with the individual (this focus, as we have seen, distinguishes it from the medieval model). It assumes that the "well-being of the whole is the *final goal*, but that this is reached through attending to the needs and desires of the many subjects that make up the community."[63] The ecological self "is constituted by relationships and exists only in relationships,"[64] an enlarged self that recognizes its connections to others, expanding its scope of concern and care to all others.[65] Simply put, the ecological model is a model of community, and the concept of community implies an ethic of care. In order to care for other persons appropriately, McFague argues, we must pay close attention to what they need. The same is true in caring for nature: love for nature must "be informed about what the natural world needs to survive and flourish."[66]

Thus the ecological model and the Christian ideal of love underscore the same ethic of love and care for all subjects as subjects. So what is it that nature *needs*? What, if anything, does loving nature require us to do? In the most basic sense, McFague argues, we are to develop greater sensitivity toward nature, gain a greater appreciation of it. In some cases, she concedes, loving nature is not entirely spontaneous or easy. The best we can manage is respect toward certain aspects of the natural world. The community ethic is not always "love and harmony"—think of poison ivy, cancer cells, the AIDS virus, she notes. How can we love these entities? The best we can achieve in these "difficult cases" is a simple acknowledgment that these forms of life are nevertheless "subjects in their own worlds" (152). McFague does not counsel us to refrain from destroying the AIDS virus. We may do so, but we should still retain a basic sensibility of respect even as we struggle against these threatening forms of life.

Fortunately, with most organisms love and care (not merely a more detached form of appreciation or respect) are possible and even warranted. Yet Christians often fail in their attempts to love other beings as they love themselves. McFague argues that Christianity contributes something unique to the ecological model in this respect: it preaches forgiveness for human failures. Christians can locate clues to what living in accordance with the ecological model would be like, if we could successfully carry out this ethic, from stories "in which the lion and the lamb, the child and the snake, lie down together; where there is food for all; where neither people nor animals are destroying one another" (158). Even if such communities do not exist on earth, McFague concedes, "the Kingdom of God and the eucharistic banquet are such clues for Christians." They represent ideals that Christians strive to achieve, even as they inevitably fall short.

The ecological model and the Christian model contribute in other positive ways to one another. The ecological model serves to remind Christians that love for other subjects is a perspective that is *not* limited to human beings. Christians are to love *all others* as they love themselves: disinterestedly, for their own sakes (164). Ecology expands the Christian's understanding of love. Just as the ecological model widens Christianity's definition of others, the Christian ethic of care adds a depth dimension to the ecological model because of Christian preference for the "oppressed, the poor, the despised, the forgotten." In other words, ecology widens the scope; Christianity sharpens the focus. Christians, therefore will focus, on the "logic of their own faith, on healing the wounds of nature and feeding its starving creatures, even as they also focus on healing and feeding its needy human beings" (169).

Examples of needy and oppressed nature toward which such an ethic might be directed include local neighborhoods degraded by toxic wastes and landfills, wilderness areas threatened with clear-cutting and loss of biodiversity, as well as the dwindling parks of urban environments. All these areas constitute nature, and all are instances of nature as the "new poor" (170).[67] These examples of "nature" in need of the Christian preferential ethic indicate that her ethic is meant to apply to both wild and nonwild nature equally—i.e., the urban environment as well as the pristine forest. However, McFague is not always clear about what sort of nature she has in mind. Given her own equivocation on this point, I will try to apply her ethic to both kinds of cases, wild and cultivated, to see how well it fares in specific examples of our encounters with nature.

APPLYING MCFAGUE'S ETHIC TO NATURE: TWO SCENARIOS

Loving Nonwild Nature

We have seen that McFague defines "nature" broadly: nature is a local spot we know and love, and it is also everything in the universe, from viruses to black holes. We engage it as a worldview as well as a daily experience. To which of these parts of nature does she intend an ethic of love and care to be extended? In some cases she writes as though she is not dealing with nature "as a whole"—i.e., wild nature, nature beyond our immediate experience—but only the nature that we encounter in our day-to-day lives, within our own neighborhoods and communities.

McFague distinguishes nature as "wilderness" from nature as "garden." Nature as garden represents cultivated nature—nature shaped in both positive and negative ways by humans. The garden constitutes a kind of "second nature," she argues, borrowing a phrase from Michael Pollan. Pollan observes that among environmentalists, particularly American environmentalists, nature has often been divided into two categories, wild nature, which makes up only about "8 percent of America's land," and cultivated nature, which is everything else (159). We have done a reasonably good job, he argues, of developing wilderness ethics in America (think of John Muir and Aldo Leopold), but have tended to ignore the rest—the vast majority of land—dismissing it as already tainted and not "true" nature.

McFague concurs with Pollan, adding that "the *real test* of an ethic for nature is how it deals with the 92 percent—the big part that is more like a garden than a wilderness" (ibid.; my emphasis). She does not mean by this to discount the remaining 8 percent however. Christians should "certainly preserve true wilderness" as well (ibid.). But, citing Pollan again, she notes that "[he] says that the ethic of care that emerges from trying to take care of the garden is different from the wilderness ethic" (ibid.). This is the first indication in *Super, Natural Christians* that *two* ethics may be necessary or that the ethic she so carefully developed is *not* meant to apply to "wild" nature. However, in the very next paragraph she states, "The ethic we need is not only one for the wilderness but also one for the garden, for the second nature that we share with the first nature" (160). Does this mean she supports a general ethic of care but that this ethic will vary in its specific application to wild and cultivated nature?

It is not entirely clear what McFague intends here. She has previously defined nature broadly enough that the definition includes all its forms. Furthermore, if her contention is that we need a separate ethic for wild nature, she never develops one. Her argument up to this point has suggested that Christians are to love *all* nature, *all other beings* as they love themselves. She stresses repeatedly the all-inclusiveness of the Christian ethic if it were applied to nature: "We would have *one way* of being, knowing, and doing in relation to God, other people, and nature" (2). An ecological, evolutionary ethic, moreover, certainly *sounds* like an ethic relevant to wild nature—i.e., ecosystems, evolving species, natural habitats—not just city parks and backyard birdfeeders.

Given the ambiguity of McFague's remarks in these passages, I will assume for the sake of argument that the ethic she proposes is intended to include cultivated and wild places. So how would we apply the subject-subjects model in our encounters with nonwild lands and animals? McFague proposes that we consider the way this model influences our thinking about other subjects that we encounter in a daily context: animals used as food. How does the subject-subjects model respond to the situation of "chickens being raised for human consumption in inhumane conditions"? (39).

Here we have a concrete example of a common interaction with nonwild nature. McFague argues that, in keeping with her model, "I might decide that the cruelty involved in commercial chicken farming is such that I will not eat chicken" (ibid.). Other options include the decision to eat "only free-range chickens or to work for legislation to improve the conditions under which chickens are raised" (ibid.). Vegetarianism is an option, but it is not required by the subject-subjects model. But why isn't vegetarianism *required* of this model? After all, it would seem that the model involves loving others as one loves oneself and other humans; this model, she has argued, involves treating others as ends in themselves, not objects to be used by us, to our own ends. It is, in McFague's terminology, a noninstrumental, nonhierarchical, nondualistic, and nonanthropocentric model. The subject-subjects model sees the world through the "loving" eye that cares for the other as other rather than the "arrogant" eye that sees the value in other things as relative only to oneself and one's own needs and desires. How can one justify eating another subject, even if raised and killed humanely?

McFague really has no answer to this question in terms that are consistent with the Christian ethic she wants to extend to nonhuman animals. Instead, she defers to Native American customs, noting that their traditions "have related to animals as subjects" but "have also condoned hunting and meat-eating" (ibid.). Perhaps so, but Native Americans are not Christians seeking to love nature in keeping with Jesus's ministry of love, healing, and caring for all subjects. Native Americans may very well have a notion of animals as subjects, but their understanding of human relationships with animals is presumably not identical to the Christian perspective McFague is explicitly promoting. In short, she never explains how a *Christian* nature spirituality, as she develops it, can justify eating animals.

The inconsistencies in McFague's account emerge when we reflect upon her discussion of different "kinds" of subjects—a chicken as opposed to an AIDS virus, for instance. We have seen that she denounces hierarchical views of nature, including the vertical grid of the medieval model. How does her own model deal with difference? Clearly, McFague does not think that we can or ought to love an AIDS virus in the same way that we might love a sentient being. The virus occupies a particular level of subjecthood, relative to other subjects such as chickens and human beings. Thus chickens raised for human consumption "present a different level of subject" than the virus.

Note that the subject-subjects model in fact assumes that there are different *levels* of subjecthood. We have no obligation to care for some and, in fact, can kill them with little or no compunction. The killing of others, such as a chicken, is somewhat more problematic though not strictly forbidden, nor does McFague indicate that we are under obligation to provide care or love for subjects at this level. We *may* decide only to eat so-called food animals (free-range subjects?) raised humanely; she implies that if the living and slaughtering conditions of such animals are improved, the ethical problem of eating them ceases to exist. In what sense does this ethic treat all beings as subjects "like us"? How does this approach support a single ethic of being, knowing, doing that McFague endorses?

McFague constructs a system of subjecthood that implicitly assumes a hierarchy of beings. One's "response would be different" to different beings because they "represent a different level of subject" (ibid.). Her approach is not nonhierarchical or nonanthropocentric. Like Ruether's failed attempt to develop a consistent ethic of biophilic mutuality, McFague's argument for loving all beings *as we love other people and God* gets lost along the way. An ethic of radical relationality is abandoned in favor of one that assumes different "levels" of being, and yet the justification for different levels of being is never provided. She presents a picture of nature that is scientifically inaccurate and then develops an ethical translation of that picture (love and care for all subjects *as* subjects) only to depart from this ethic when it demands something radically different in our treatment of nonhuman animals. McFague's radical reorientation turns out to generate a surprisingly conservative ethic.

I have attempted to run McFague's ethical "experiment" with regard to one particular example of nonwild nature. Assuming, as I have for the moment, that she intends her model to apply to nonwild nature, some problems immediately become apparent. Other aspects of her ethic pose even more serious problems when applied to wild nature and animals.

Loving Nature as Wilderness

Despite this detailed discussion of cultivated nature, I do not think McFague intends her ethic to apply primarily to backyards, city parks, or domesticated animals. As we have seen, she describes the ministry of Jesus as illustrated in his feeding and healing of the needy and then entreats Christians to model their behavior toward nature on his praxis. At first glance many people would not necessarily see this ethic as counterintuitive or particularly strenuous. After all, most of us have had the experience of caring for and feeding some representatives of "nature" and, in doing so, have often sensed some kind of reciprocity in our relationship with them. Those of us who have pets, birdfeeders, houseplants, or even just city parks with pigeons experience this regularly. Given that McFague understands the requirements of Christian nature spirituality to be "radical" and "counter-cultural," presumably she does not merely have in mind feeding and healing domesticated animals, something many people already do without hesitation. In applying the radical, destabilizing praxis of Jesus to nature, McFague must mean that we are to extend this ethic in all directions, particularly in those directions that are novel and *counter*intuitive. Recall that the emphasis on Jesus's ministry provides a link to the inherited Christian tradition, while the extension of this ethic to nature constitutes a break with that tradition, the conceptual shock that is necessary if we are to alter basic patterns of thought and behavior. This, as we have seen, is the way that McFague's experimental theology typically proceeds.

Caring for nonwild nature cannot be all that McFague is asking of us. So what would it mean to extend this praxis to wild nature? A very basic review of Darwin's account of natural selection would remind us that evolution occurs by means of differential survival and reproductive rates among animals engaged in competitive (as well as cooperative) interactions. Animals vie for limited resources, both in terms of food and mates. The number of animals in a given population generally exceeds the amount of resources available for sustaining the population. Those organisms that are, for one reason or another, better adapted to the particular vicissitudes of their environment will survive and (more important) reproduce.[68] The death of some organisms—sometimes a large percentage of them—by starvation, disease, or predation maintains relative fitness within that particular context and checks their numbers.

This sense of checks and balances, of natural limits, is an idea that environmentalists—including McFague and Ruether—frequently emphasize in their descriptions of the interdependence of ecosystem communities. What they often fail to take seriously is that these limits are maintained at great cost to *individual* animal lives. The ecological community—assuming that "community" is the appropriate term for natural systems—does not aim toward the good of each individual within that community, as (ideally) human communities do. In *Life Abundant* McFague again stresses that God's (and the loving Christian's) vision for earth is a kind of eucharistic household (*oikos*) in which, unlike the world we now inhabit, "everyone does eat," where there is an "equitable sharing of resources" for all God's creatures.[69] She proposes a "God of the basics" who concerns himself "with those things that keep beings alive, healthy, and flourishing."[70] But while a human model of community may entail care for the health and well-being of each individual, nature does not. Evolution is not simply a cooperative venture. However well-intentioned, McFague's endorsement of interventions in nature to feed its starving creatures and heal their wounds would be counterproductive and perhaps even disastrous in light of natural selection.

Again, McFague may well object that I am taking her account of Jesus's praxis of healing and feeding too literally, that her theology deals in metaphors. Perhaps the example of Jesus's ministry is meant to be suggestive and metaphorical as well, a way of pushing beyond prejudices toward nature that we all hold, consciously or unconsciously. We have seen that Ruether too sometimes understands healing in this way, as a means of overcoming conceptual dualisms and bridging gaps in our consciousness. Perhaps McFague also has this sort of healing in mind rather than actually attending to the physical wounds of animals. Feeding too might be intended as a metaphor for caring, giving, and sharing. It expresses an openness toward others, despite their otherness. We invite others into our homes, metaphorically speaking, into the very centers of our existence, give them a place at our tables and nurture them. In *Life Abundant* symbolic inferences are likewise drawn from the Lord's Supper as a "banquet where all are welcome."[71]

Yet, McFague is quite clear in *Super, Natural Christians*, as well as in her theology as a whole, that Jesus's ministry highlights the importance of bodily life and well-being. As she stresses many times, Christianity has a tradition of valuing individual *bodies.* Jesus's ethic proposes that bodily health is a basic right. He did not just minister to outcasts *spiritually;* he cared for them *bodily*, and he directed his attention to those who were most in need. He acknowledged that bodies are important; he himself was embodied. McFague argues that Jesus's praxis implies that "all bodies need to be fed."[72]

McFague interprets Jesus's ministry as an expression of the overarching desire of God for the "well-being of not just humans but of all creation."[73] Like Ruether, she argues that health involves redressing the imbalances in the natural world. But in nature, the health and wholeness of *all beings* simply cannot be maintained simultaneously. Some beings survive at the expense of others.[74] All bodies may *need* to be fed, but this need is not met for all in nature. Physical health is not a basic right in the natural world, nor is it the appropriate role of humans to see that this need is met. In fact, wildlife management (which, in any case, operates at the level of *species* and not just individuals) often has to take a triage approach, directing attention away from the "neediest" and most imperiled organisms, allowing limited resources to be applied to preserving species that are more likely to survive. As with war, Holmes Rolston argues, so it is with wildlife strategies: "One abandons the most desperately injured, cruel though this may seem, because they will probably not survive in any event."[75]

McFague's ethic falsely imagines that nature functions in a way that permits the flourishing of every individual creature at once. As Northcott points out, McFague sidesteps "discussion of the real moral conflicts which living together in a physical universe" necessitates, owing to her commitment to "dialectical pantheism" and vague egalitarianism, both of which render ethical decision making virtually impossible.[76] As in Ruether's arguments, there is a basic inconsistency between the emphasis on interdependence as central to the ecological perspective, on the one hand, and the role prescribed for humans as healers and sustainers of life, on the other. Of course, McFague concedes that humans seldom successfully carry out this praxis in regard to nature, but she seems to believe that the flaw lies not with the ethic itself but with human flaws that prevent us from doing the right thing, even when we know what that is. Thus she presents the image of a banquet when all creation is healed and fed as a symbol of hope, an ideal that Christians continually strive to attain on earth.

As an ideal for nature, however, this hope is misplaced: drawing on biblical passages, McFague describes the eschaton as a state in which there is food for all creatures, a time in which predators and prey lie down together and no creature will ever again destroy another. These images, she argues, suggest what living in a "true" community would be like. The "community model" (which, as McFague has already established, corresponds to the ecological model) "attend[s] to the needs and desires of the many subjects that make up the community."[77] By attending to the bodily needs of individual animals, McFague hopes to *create* an ecological community that does not actually exist on earth.

Here we return to McFague's confused understanding of models and reality. The ecological model has less to do with a scientific picture of reality than it does with a biblical, eschatological hope for the future. McFague's invocation of the eschaton makes explicit something that has lurked beneath the surface of her argument all along: the form of ecology she has in mind does not really exist anywhere on earth. This is why she is sometimes led by her own argument to speak, paradoxically, of extending the ecological model *to nature:* "How," she asks, "can we apply the ecological model to nature, and how can we develop the loving eye that respects nature as subject?"[78] The Bible holds some clues, she argues. Thus, it turns out, the ecological model is the unattainable ideal. The "reality" is something different altogether. McFague's agenda now becomes clearer: what her environmental ethic seeks to do is *not* to derive ethical guidelines from an ecological model rooted in reality but, rather, to *transform* reality by means of a biblical model. The ecological model is not the "modern evolutionary, ecological perspective." It is simply the peaceable kingdom, a corrective to nature as it really is.[79]

But an environmental ethic must be rooted in biological realities. We cannot hope to change nature by engaging it as though it were, or could become, a perfect ecological community. McFague's ecological model and the ethic it supports do not meet the criteria for choice of metaphors and models that she defines as important: 1. this model is not adequately informed by science and 2. it is not effective in the context in which it must be applied, whether that context is wild nature or places and animals cultivated by humans.[80] The ecological/subject-subjects model fails to meet her own standard of "truth," which is defined, pragmatically and practically, as effectiveness in context. On the whole, McFague sidesteps any serious discussion of evolutionary theory in her discussions of scientific reality, in her historical sketch of paradigm shifts in Western interpretations of nature, and in the fading language of "evolution" that characterizes her account of the contemporary scientific view of nature. Like Ruether, she is much more interested in (often questionable or outdated) ecological themes than in the evolutionary processes that might cast doubt on the compatibility of the ecological ethic with a Christian model of love and care for earth others.

We will conclude with an examination of a theologian who discerns some of these flaws in ecotheology, yet remains drawn to many of the communal and covenantal themes that are expressed in Ruether and McFague's writings. As noted previously, Michael Northcott takes issue with many of the claims of ecotheology, particularly when attempts to value "otherness" come at the expense of practical applicability in environmental ethics. He is nevertheless sympathetic to ecocentric orientations and he applauds feminist perspectives for their emphasis on relationality and ethics of care. "Feminist approaches to ethics point to the stuntedness of ethical discourse and behavior in patterns of relationality, intimacy, connectivity, community and care" rather "than atomistic individualism and self-interest."[81] Northcott agrees with ecofeminists that such approaches are especially fruitful for overcoming our alienation from nature.

He seeks to construct a Christian ethic consistent with some dimensions of ecocentric and relational ethics, but at the same time he is concerned about certain tendencies in many Christian pronouncements regarding ecology: often, in their attempts to mitigate the anthropocentrism of Christian theology, they attempt a too radical decentralization of humans, or they tend to collapse differences between organisms into sameness, homogenizing all life-forms. Homogenization makes it difficult, if not impossible, to sort out conflicts of interest or clashes of value in environmental ethics. In this context he criticizes McFague for annihilating all differences between life-forms, "conceiving the whole, and its various parts, as extensions of the God/self to which we all belong" (158). He correctly concludes that her approach offers "few practical glimpses of how we might actually discern between the needs of different species and organisms, given her intrinsic valuation of every living and even inanimate particular as God-breathed and inspired" (159). Significantly, he notes, her ethic fails to offer a "clear account of how we might understand natural evils such as predation" (ibid.).

Northcott poses a set of questions that get to the heart of McFague's confusion about the connection between Christian theology and evolutionary—and social—realities: "Why," he asks, "should a monistic cosmos which is identified with God lead us to respect the vulnerable more than the powerful?" (ibid.). If God is identified with "life" in *all* its forms, what grounds do we have for assuming that God is on the side of the oppressed rather than the oppressors? Interestingly (quite contrary to my own interpretation), Northcott finds fault with McFague's arguments on the grounds that her science is essentially *too* modern, rather than too antiquated. He attributes McFague's shortcomings to her (and other theologians') adherence to a picture of creation "as described by modern science"—an order of nature so "deeply morally flawed that we dare not read the mind of God from it" (ibid.). In other words, McFague mistakenly and too readily accepts "an anti-teleological world-view" as a prerequisite for any attempt to engage science and theism. The modern picture of nature is too morally bankrupt to provide any meaningful guidance, so Northcott rejects this account of creation, seeking a "more precise ordering of human and non-human goods and values" that can, in some sense, be read off the world around us (163). He is considerably more impressed with Ruether's arguments, which he considers "humanocentric" insofar as her ecological ethic is filtered through the experiences and claims of women as an oppressed group. Indeed, Ruether and Northcott share an emphasis on covenantal community, and both understand cooperation to be the overriding theme in the natural world.

On many points I agree with Northcott's characterizations of McFague and Ruether, though I would add that McFague's ethic is simultaneously homogenizing *and* hierarchizing (that is, her homogenizing move unwittingly retains hierarchical distinctions: organisms are subjects like us and those most like us are worth more.) At the same time, I contend that his interpretation of science and ethics suffers from many of the same flaws as other ecotheologians. Northcott displays a similarly selective use of science that ultimately skews his ecological ethic in the direction of harmony, balance, and mutuality. In fact, he likewise turns to the biblical image of the peaceable kingdom as a guiding light for proper ecological relations.

Drawing upon the covenantal worldview of the Hebrew Bible, natural law and virtue ethics, and inferences from ecology, Northcott proposes an alternative that resists homogenizing and monistic tendencies of ecotheology. He is particularly interested in establishing that the created order of the world contains within it a moral order. The way the universe is, he argues, determines how we ought to behave in it (165). Discerning (or perhaps rediscovering) that moral order is key for understanding how we are to treat nature and nonhuman life; developments in science play both a positive and negative role in this process of discernment. Science is in part to blame for stripping nature of its moral order: like many ecotheologians and secular ethicists, Northcott laments the destruction wreaked by "Cartesian," mechanistic, and atomistic approaches that drained the natural world of life and intrinsic value. But he is far less prepared than some to dismiss the scientific method, to condemn the empirical, objective, "arrogant" eye (as McFague calls it) of the scientific observer. He argues, for example, that

> it is precisely the modern scientific method of empirical observation, applied with true ethological consistency to animal sociality and organic connections in the nonhuman world, which allows us to dare to believe that the nonhuman world, as well as human society, still contains within it marks of the moral order, as well as the physical design, of the creator God who first affirmed its goodness. (132)

If science has at times buried nature's moral order, it can also provide the tools to unearth it once again.

A biblical covenant ideal is central in Northcott's theology, and it plays essentially the same role in his ethic that the term *community* plays for other ecotheologians. Covenant similarly implies that humans are to envision themselves as part of a community of life whose members are human and nonhuman. Northcott prefers the covenant metaphor because it more emphatically includes God as the originator and sustainer of an ethical relation, affirming that "the world belongs to God and so human power is only to be exercised in relation to God's purposes for justice and peace in the world" (130). Covenant embodies God's "particular love for the poor and vulnerable, including children. . . . God is opposed to all forms of racism, sexism, and classism" (ibid.). With regard to the natural world, certain features of the covenant idea are especially salient. These include preservation of the "integrity and intrinsic value of ecosystems," a rejection of the treatment of nature as mere "resources for human exploitation," and, most significantly, the link that covenant establishes between nature's integrity and human societies (ibid.). "Social balance and ecological balance," he writes, "are interrelated" (196).

In delineating the crucial features of the covenantal ideal, Northcott begins to sound strikingly similar to some of the ecotheologians of whom he is otherwise critical. The same questions arise regarding the relationship between his use of scientific concepts such as "ecosystems" and theological concepts such as "covenant." In precisely what sense is the "order" and "balance" of human societies linked to the "balance" of natural systems? On one level Northcott is merely pointing out that human moral choices and social structures—the pursuit of peace rather than war, the establishment of just societies rather than unjust ones—have a huge impact on the environment. It is undeniably the case that degradation of the environment and human poverty and oppression are intertwined. But Northcott's claims go deeper than this, and their connection to ecological realities becomes strained when he explains what he means by the moral, created order. Northcott clearly believes that the created order has been corrupted by sin and that ecological responsibility involves restoring nature to a pre-Fall condition.

Northcott's interpretation of natural systems is premised on a conviction that elements of chaos, imbalance, instability, and violence are the result of human sin, past and present. In ignoring the relationality of nature, he argues, we run the risk of "bringing chaos into the biosphere" and inviting "unpredictability" (165, 196). Indeed, he defines covenant in contrast to ecological chaos, competition, unpredictability, and disturbance, and in doing so he aligns himself with ecotheologians who interpret community as an antidote to all negativities in nature. For example, he alludes to biblical stories (such as Jeremiah), tracing the source of ecological upheaval and environmental stress to abandonment of covenant responsibilities. Covenant coincides with the "imposition of order" on the "the raging forces of the cosmos" (167). Disruptions of nature "are signs of disturbances in human and divine relationships, of human mismanagement of nature and society, rather than evidence of natural evil" (197). The covenant metaphor invites us to "construct a society where domination, inequality and competition are challenged and resisted, both in relation to their impacts on human flourishing and on the flourishing of the nonhuman world" (179).

Northcott shares other ecotheologians' deemphasis of all Darwinian elements of nature and animal behavior. He acknowledges the existence of more competitive views of nature, including human nature (and here he cites Richard Dawkins, whose work he associates with a "survival of the fittest" stance), but he claims that these are modern myths that exaggerate competitive, violent tendencies, and "justify ethical atomism" (175). He notes with approval that Donna Haraway and other "women primatologists" see far more "co-operation, mutuality and altruism" than competition and conflict (ibid.). Humans have turned away from this moral order in their own communities and have compromised it in the natural world. In a passage that echoes Ruether's account of human competitiveness as deviation from natural, cooperative norms, Northcott concludes that "co-operation, mutuality and altruism are indeed part of the natural order which humans, more than any other animals, have successfully managed to distort and eschew." "Competition for scarce resources," he contends, is "not the natural condition of life" (196). On the contrary, nature's genuine (and original) goodness shines through in cooperative, mutualistic behaviors—these are remnants of a moral order now corrupted as well as signs of hope for an order yet to be restored.

For Northcott, as for McFague, the ideal ecosystem is bound up with biblical images of the peaceable kingdom, of harmony and justice. The Hebrew Bible presents us with an ecological ethic rooted in a "shalom of the earth"—a "peaceable kingdom of shalom and ecological harmony" in which "predatorial behavior will no longer characterize human and non-human relations" (194). The New Testament materials depicting the crucifixion and resurrection of Jesus offer fulfillment of the Hebrew prophets' hope that the world will be redeemed from suffering and death. Jesus's triumph over death has implications for all life-forms, for an "ultimate transformation of created order" in which "enmity and violence will be no more" (202). Only from the standpoint of resurrection, Northcott argues, can we see "creation truly for what it is," namely, the "product of the holy will of God whose plan is to restore its goodness, wholeness and harmony" (ibid.).

It is not clear how Northcott's ethic avoids the pitfalls he identifies in other ecotheologies. Doesn't his claim that the *entire created order*, human and nonhuman alike, cries out for redemption, harmony, and peace constitute a grand homogenization, a collapse of important differences between myriad forms of life? He is critical of McFague's inability to offer practical solutions, the failure of her ethic to recognize the differing needs, as well as conflicts of interest, among life-forms. Yet an environmental ethic that seeks harmonious and peaceful relations among all beings surely cannot take seriously the particular needs, the specific ways of life, of animals—take, for example the needs of predators, whose means of survival will apparently be revoked when the original goodness of creation is restored. The various life-forms may have their own individual needs, drives, and purposes, but ultimately God's plan for creation eliminates those differences, as the "frustration of every living thing is drawn into the eternal purposes of God for the restoration of the cosmos" (204). Northcott's desire for peace, harmony, and justice in the natural realm constitutes a human agenda imposed on nature as much as (if not more than) Ruether's efforts to understand nature through the lens of women's oppression. True, Northcott does retain a privileged status for humans (and thus a hierarchical structure) that is more pronounced than some ecotheological arguments, which he deems too "de-centralized."

But, in many other respects, his arguments fall in line with Ruether and McFague's.

What becomes of the empirical, objective scientific methodology that Northcott regards as a valuable tool for rediscovering an order in nature? Apparently the tool is of interest only when it uncovers cooperation, altruism, harmony, and relationality. Where science finds scarcity, competitiveness, or conflict, it is merely perpetuating unfounded human myths about, and sinful distortions of, the "true" moral order, the created world as it "really" is. Where science does not provide the teleological and harmonious narrative that Northcott seeks, he simply revives an older narrative or invents a new one. Such is the case with his definitions of evolution and ecology, both of which affirm nature's goal-seeking moral order. The "principle of natural selection," he insists, "is teleological because it describes intention, purpose" (249). The drive to survive necessarily implies a *telos* (here Northcott, like other ecotheologians, mistakenly equates Darwin's theory with simple "survival"). "But why do beings want to survive?" he demands. Any logical answer, he believes, entails commitment to teleology. Modern scientists' denial of teleology is shown to be groundless, Northcott maintains, by the testimony of their peers in the discipline of *ecology* as well. In this context he takes the well-worn path of many ecotheologians, citing the authority of Gaia scientist James Lovelock, as well as Eugene Odum, the ecosystem ecologist who popularized the organismal, equilibrium account of ecosystems in the 1950s. He thus concludes that "ecosystems, like animals pursue certain goods after which they are teleologically ordered by the creator" (253). Ecosystems (undisturbed by humans) adhere to the same covenantal ordering that God originally imposed on nature. An ecological ethic involves seeking to "preserve and restore the stability, harmony, and relationality of all things" in keeping with the "wisdom of natural systems" (197). In short, he shares Ruether and McFague's romanticized understanding of evolution and ecology, but his approach is distinct in its explicit and conscious rejection of what he perceives to be "postmodern" science and its implications.

Though he does not clearly avoid the pitfall himself, Northcott's identification of a homogenizing trend in ecotheology (and the problems it raises) is one of his most useful contributions to the discussion of Christianity and environmental ethics. We have seen McFague (and to a lesser extent Ruether) insisting that we view all life-forms as subjects like us, seeking to flourish within the community of life. In fact, the emphasis on seeing nonhuman animals as humanlike subjects is one that frequently recurs, both in religious and secular environmental ethics. Just as interdependence is interpreted as self-evidently good, "discrimination" among organisms as subjects is often assumed to be unethical in ecological theology.

In the next chapter we will turn to a discussion of ecotheologians Charles Birch, John Cobb, Jürgen Moltmann, and, to a lesser extent, elements of Larry Rasmussen's approach. The ecological model and its corresponding treatment of animals as subjects plays an important role in Birch and Cobb's theology; unlike McFague, however, they do not extend a single ethic to all beings but present a scale of sentience, or what they call "richness of experience," as a criterion for discerning our obligations to different classes of nonhuman animals. While there are advantages to an ethic that discriminates between different kinds of beings, Birch and Cobb's basis for discrimination is problematic, as we will see. In chapter 4 we will look at another variation on the theme of animals as subjects in the secular arguments of Tom Regan and Peter Singer. In urging a radical reconsideration of our treatment of nonhuman life, Regan and Singer present arguments similar to McFague's. Here, however, a radical reorientation is accompanied by an *equally* radical ethic. Indeed, these arguments are considered too radical for many, and Regan and (especially) Singer are often dismissed as extremists.

For now, however, we will continue with a critical examination of ecological theologians who are concerned with liberating life from conditions of oppression and suffering, and from mechanistic modes of thought, on the grounds that liberation is indicated by both biological and religious considerations. The community model of nature (or a single community of humans and nature) is promoted as a key ecological insight in these arguments, and, once again, evolutionary processes are either omitted or reinterpreted in light of this ideal.

CHAPTER 3

The Ecological Model and the Reanimation of Nature

The ecology movement has reawakened interest in the values and concepts associated historically with the premodern organic world. The ecological model and its associated ethics make possible a fresh and critical interpretation of the rise of modern science in the crucial period when our cosmos ceased to be viewed as an organism and became instead a machine. The removal of animistic, organic assumptions about the cosmos constituted the death of nature.

—Carolyn Merchant, *The Death of Nature*

If organisms are seen as mechanisms, they will be treated as such, and as such we will treat each other. The very concept of health, of wholeness, disappears, just as organisms have done from modern biology. A biology of parts becomes a medicine of spare parts and organisms become aggregates of genetic and molecular bits with which we tinker as we please, seeing their worth entirely in terms of their results, not in their beings. This is the path of ecological and social destruction.

—Brian Goodwin, *How the Leopard Changed Its Spots*

The conviction that mechanical views of nature have wreaked environmental havoc is woven throughout the writings of ecological theologians. If science itself is partly to blame for oppressive attitudes toward life, then perhaps a new and better science—enlightened by theological insights—can pave the way toward a more ethical treatment of all creatures. The ecological model, as the product of scientific and theological reflection, promises to provide this fresh outlook on life. We have seen arguments in favor of the ecological model already in the work of ecofeminists. Their support of the ecological model is shared by Christian environmentalists who approach nature from slightly different theological perspectives. In this chapter we will examine the work of ecotheologians who ground their understanding of life in eschatological theology (Jürgen Moltmann) and process thought (John Cobb and Charles Birch).

In the discussion of Moltmann, Cobb, and Birch that follows, I will not attempt a survey of all or even most of their works, but will focus on those texts in which they present what they see as the main points of convergence between their theological presuppositions and environmental issues. As in previous chapters, we will pay particular attention to ecotheologians' understanding of ecology, and interdependence, and consider whether their use of those terms is consistent with Darwinian views of nature. The idea of liberating nature from a state of oppression, suffering, and fallenness will also form a secondary theme. Specifically, we will look at the problematic relationship between the objective of eradicating suffering and an evolutionary understanding of nature.

Like McFague and Ruether, these authors attempt to develop an ecological theology that relies on modern scientific insights about nature and its processes. They similarly propose that the ecological model supports ethical treatment of nature and its many life-forms. Again, this model is rooted in the idea of community and mutual interdependence, and it is often presented—as we saw with ecofeminists—as an insight that is both derived from nature and applicable to it. In the work of all these ecotheologians the ecological model is consistently assumed to provide an alternative to the mechanistic, Cartesian philosophy of science that objectifies nature as dead matter. In their eagerness to move science away from the older mechanistic model, toward the "modern" organismic, ecological model, however, some ecotheologians tend to gloss over Darwinian processes in nature. Instead they embrace a type of theistic evolution that is not representative of Darwin's theory but provides a better fit with the theological orientations they espouse. The work of Moltmann, Cobb, and Birch illustrates this trend particularly well.

These three authors share a commitment to articulating the role of God in natural processes. For Moltmann, God, or the spirit of God, plays a direct, guiding role in evolution; for Birch and Cobb, God acts more indirectly, "luring" nature forward in the midst of infinite possibilities. These two different understandings of God support rather different interpretations of the presence of evil and suffering in nature. In positing a direct involvement for God in "creation," Moltmann tends to deemphasize evil and suffering, interpreting nature as predominantly harmonious. His emphatic affirmation of the harmony and perfect adaptation in nature surpasses even the Romantic ecologies of ecofeminists such as Ruether. At the same time, his work exhibits a similar ambivalence about the natural world; Moltmann is unsure how much suffering occurs in nature and calls for its liberation from oppression in spite of his belief in its overriding harmony.

Birch and Cobb's theology openly addresses evolutionary processes, and the conflict and suffering they generate, depicting God as responsible but not indictable for these problems. Despite what they see as the inextricability of suffering from "enjoyment" in all life-forms, they propose an ethic that aims at the reduction or removal of suffering. It is not clear, in either Moltmann's or Birch and Cobb's theologies, why the paradigm of liberation from suffering is appropriate, given their own interpretations of evolution. Moreover, as we will see, Birch and Cobb present ethical guidelines for our treatment of all life that, like McFague's ethic, could not be applied to the natural world without causing serious disruption to the very processes of evolution and ecology they claim to take seriously.

We will begin by considering the basic features of Moltmann's ecological theology and then turn to the work of John Cobb and Charles Birch (both their collaborative efforts and individual contributions).

MOLTMANN: *GOD IN CREATION*

The main points of Moltmann's ecological theology consist of his trinitarian doctrine of God, a panentheistic interpretation of God's relationship to nature, a basic eschatological and messianic orientation, and his understanding of the Sabbath symbol as a promise of eschatological peace and liberation for all creation. Let us look briefly at each of these features and then consider their implications for environmental ethics and Darwinian interpretations of nature.

Moltmann subscribes to a triune doctrine of God that explicitly rejects the separation of God from creation. The concept of a triune God, he argues, helps to dispel the myth of our unilateral relationship of dominion over nature. Like McFague, Moltmann proposes that we adopt a "non-hierarchical, decentralized, confederate theology."[1] According to this account, the third part of the trinity—the spirit of God—is crucial for understanding God's role in creation. Creation is "creation in the spirit"; God's spirit dwells in creation, preserving and renewing all living things. The trinitarian view is also panentheistic in orientation: God created the world and the world comprises part of God's being. Because of the triune nature of God, God can be understood without contradiction as immanent in nature and transcending it.

Moltmann's understanding of the spirit is closely linked with a community interpretation of nature, for it is the spirit that "fosters community" and maintains the "warp and weft of interrelationships."[2] Working within the processes of evolution, the spirit exhibits a "recognizable tendency to develop into more and more complex open systems."[3] At every evolutionary stage the spirit "creates interactions" among all living things, and these interactions culminate in an overriding harmony of nature. Divine spirit represents what Moltmann considers the "only realistic alternative" to death and destruction for the future of creation, namely, a "non-violent, peaceful, ecological world-wide community in solidarity."[4] In other words, the possibility of hope for the future of the earth is inseparable from a belief in the immanent creator spirit.

Moltmann's ecological objective of sustaining communities of solidarity is a recurring, but scientifically dubious, theme in his theology. We will see that his interpretation of ecology not only leaves out important evolutionary processes but in fact renders natural selection superfluous. The problem lies in Moltmann's inappropriate selection (and rejection) of certain scientific information. As one critic of ecotheology has observed, environmentalists are often deeply ambivalent about science, both critiquing and embracing it as suits their purposes: ecotheology "too often simultaneously attacks and worships science and its findings," selectively invoking some scientific data while failing to recognize theories "that do not fit its holistic, steady state model of the universe."[5] Moltmann's work is a case in point: he condemns scientific accounts that objectify nature while eagerly promoting ecological "science" that supports his own brand of theistic evolution.

We have seen that ecotheology consistently attacks the "mechanistic" or "Cartesian" model that separates humans from the nonhuman world, treats animals as objects, and denies the continuity of all life. Ecotheology (most notably McFague's rejection of the "subject-objects" model) proposes replacing this perspective with one that regards all entities as interrelated subjects. Moltmann likewise disparages the mechanistic perspective, asserting the superiority of his own understanding of science. He sounds a familiar note in urging us to dispense with "analytical thinking, with its distinctions between subject and object."[6] The ecological model proposes instead that all things can be known much better "if they are seen in their relationships and coordinations with their particular environments and surroundings."[7] Ecological interdependence and the doctrine of a triune God are consistent, Moltmann argues, in that both concepts promote a *relational* view: God is not separate from nature, nature is not separate from humans. Science as a whole must break free from the pattern of desiring "to learn in order to dominate." Here the science of ecology can help, by demonstrating the "mutual relativity of human beings and the world."[8] Thus the ecological model represents an important step toward liberating science from its own biases toward life. We will see this theme again in the work of Birch and Cobb.

While Moltmann's theology shares these features (the community ideal, the critique of mechanistic science) in common with other ecological theologians, his approach places greater emphasis on the Sabbath as the key biblical symbol of hope for the earth and its life-forms. A creation theology, he argues, must be "eschatologically orientated" [*sic*]: "Faith in the resurrection is therefore the Christian form of belief in creation."[9] The eschatological vision he proposes draws on the Sabbath as a time of peace and repose for all creation. Christians have focused too exclusively on the biblical account of six days of creation while overlooking the importance of the seventh day as the true "crown of creation." Judaism recognizes and celebrates the importance of the Sabbath, he argues, and Christians should begin to do the same.[10]

As the creator's day of rest, the Sabbath offers hope for the restoration of creation as well, a time when all creation will be perfected and restored. As such, the Sabbath is also the "feast of redemption" and a symbol of liberation: God's creatures "never find the peace of sabbath in God's presence unless they find liberation from dependency and repression, inhumanity and godlessness."[11] The Exodus theme of liberation and the Sabbath promise of rest are intertwined conceptually. Weekly cycles of rest foreshadow ultimate peace for all life.

Moltmann combines the Jewish recognition of the significance of the Sabbath with a Christian, messianic eschatology. The Sabbath time of rest will bring about the "liberation of men and women, peace with nature, and the redemption of the community of human beings and nature from negative powers, and from the forces of death" (5). All relationships—those between God, human beings, and nature—"lose their tension and are resolved into peace and repose" (ibid.). The weekly Sabbath has ecological implications for our habits and routines as well. Christians should understand the Sabbath as an "ecological day of rest"—a day "when we leave our cars at home, so that nature too can celebrate its sabbath" (296).

According to Moltmann's view of eschatology, the eschatological Sabbath represents another stage in creation rather than the end of creation. He distinguishes three important kinds of creation: creation in its original form (*creatio originalis*), continued creation (*creatio continua*), and the restored and renewed creation ushered in by Sabbath rest (*creatio nova*). All three stages are part of God's ongoing creation. Consequently "even creation in the beginning already points beyond salvation history towards its own perfected completion in the glory of God" (56).

With these features of Moltmann's theology in mind, let us see whether it is possible to reconcile his account of nature with evolutionary science. How does Moltmann harmonize this view of creation with an evolutionary perspective? What does renewed creation mean for natural processes?

MOLTMANN'S THEOLOGICAL EVOLUTIONARY PERSPECTIVE

Like other ecotheologians we have discussed, Moltmann takes seriously the relevance of (certain) claims of science for theology and urges others to do the same. He earnestly seeks to bring theology and modern biological theories together in a single approach that eschews a pernicious, mechanistic conception of life. His entire theology is built around the conviction that we have to "stop thinking of creation and evolution as opposing concepts" (19). Theology and science need to cast off the "old anthropocentric picture of the world" that presents humans as the crown of creation. Until we purge our thinking of anthropocentric biases, we will never find "an adequate theological perspective for evolutionary theory" (196–200). Like McFague, he maintains that science points to an underlying picture of reality that theologians ignore at their own—and the earth's—peril. His task is to articulate the features of science and theology that contribute to a single more or less coherent vision of God, humans, and the natural world. Our global situation lends a certain urgency to this task. Given our environmental crisis, science and theology, he argues, "cannot afford to divide up the one, single reality" (34).

Despite these laudable goals, Moltmann has in fact tailored certain scientific statements about nature to make them fit the account of creation and redemption central to his theology. His work relies heavily on an assumption of harmony in nature, owing to his trinitarian interpretation of God as the spirit that dwells in creation. For Moltmann, the spirit of God *is* "the principle of evolution" (100). At every stage in the evolutionary process the spirit "creates interactions, harmony in these interactions, mutual perichoreses, and therefore a life of co-operation and community" (ibid.). Despite Moltmann's express desire to bring together a theology of creation with a modern evolutionary account of nature, his theology selects only those aspects of nature that complement a belief in harmony and community. His adaptationist account of evolution and what he calls an "ecology of space" serve to illustrate this point.

EVOLUTION AND THE ECOLOGY OF SPACE

Moltmann's ecological concept of space essentially resembles a pre-Darwinian economy of nature in which every living thing occupies a particular spot and displays a near perfect fit to its given location within that larger economy. This concept of space demonstrates that "every living thing has its own world in which to live, a world to which it is adapted and which suits it" (147). The perfect fit between organisms and their environment is disrupted by human interference and objectification of living things, expressed, again, in terms of the mechanistic, Cartesian attitude toward nature. "The Cartesian objectification of the world destroys the natural environments of living things in order to incorporate them in the environment of the dominating human subject" (ibid.). As we saw in chapter 1, Linnaeus likewise believed that every creature has a place allotted to it, a place "which is both its location in space and its function or work in the general economy."[12] Each organism's occupation of a specific space guaranteed its own food supply and other necessary resources, thus preventing conflict with other organisms and safeguarding abundance for all creatures. This divinely instituted division of labor would foster more harmonious interrelationship. By giving each organism its own place and resources, Linneaus argued, God "set up an enduring community of peaceful coexistence."[13]

Moltmann's ecology of space similarly provides security against strife and scarcity and, as we will see in a moment, leaves little for natural selection to do. Ecological spaces constitute an "environment to which a particular life is related, because it accords that life the conditions in which it can live."[14] The psalmist, he argues, describes precisely this concept of created spaces, citing the "'basic equipment'" given to the separate sectors of life, in keeping with their elemental needs (148). In short, the ecology of space is the work of "God who has made the world through his wisdom, and keeps it in existence through his Spirit" (150).

Ironically, the perfect-fit interpretation of ecology does not quite apply to humans, despite Moltmann's desire to include all life in a community of solidarity: the ecological concept of space "has developed out of research into the environment of animals," Moltmann writes, and "can only be applied to the relationships of human beings to the world within certain limits" (150–51). This follows from the fact that humans are "unique" in their ability to move out of the particular environments in which they were created. Put in theological terms, this difference between humans and animals is an expression of our "creation in the image of God." Humans are "open to the world, beyond their particular environment" (151). In our correspondence to God, who is the creator of all environments, we embody the "quintessence of all the environments of living things" (ibid.). Animals, by contrast, are created for a certain type of environment and remain more or less within those allotted places to which they were originally relegated and to which they are so well adapted.

There are several important things to note about Moltmann's argument here. First of all, he is at pains to develop an ecological model of radical relationality, yet ultimately excludes humans from the model on account of their divinely bestowed uniqueness—a move that contradicts the whole thrust of the ecological model as an interdependent community of all things. (This is a pattern we have seen before in the ecotheologies of McFague and Ruether). Moltmann's view of natural processes is un-Darwinian in any case, given its inordinate emphasis on harmony and adaptation, and is made even more so in its privileged position for humans as God's "proxy" (190). Moltmann attempts to mitigate this separation of humans from nature by noting that, while Genesis clearly appoints humans as the "apex of created things," we are nevertheless "dependent upon" animals and share with them an "animated embodiment" (187).

Yet the distinction between animals and humans that is preserved in Moltmann's account does *not* constitute a radical departure from a Cartesian, objectifying perspective; it is merely a neo-Cartesian argument for the special status of humans. Let us look more closely at his argument here. Drawing on the work of Max Scheler, Moltmann explains that what distinguishes the human from animals is "power of reflection and the imagination of the spirit." By virtue of these unique capacities, the human being "shakes off the spell of the environment, as it were, and perceives 'the world'" (150). Unlike animals, the human is capable of "interrupting his instinctive reactions and acting deliberately" (ibid.). What, then, are the markers of our interrelation with the nonhuman world? These are found primarily in our common "animated embodiment" that links humans with animals (189). Here Moltmann articulates the classic statement of the Cartesian split between humans' rationality (reason, reflection, perception) and animals' nonrational bodily instinct. Humans have a two-tiered existence: we participate in certain features of our animal nature and share with animals the things that accompany and support embodiment ("living space," "food," and "fertility"). Human beings are built up of a simpler animal nature, but they possess something more. Translated theologically, this something more is the uniqueness conferred "by virtue of [our] creation in the image of God" (151). Moltmann ascribes a "double function" to humans as *imago Dei* and *imago mundi*, and in so doing offers only a slight variation on traditional Cartesian dualism that separates humans from other animals:

> The complex system "human being" contains within itself all simpler systems in the evolution of life, because it is out of these that the human being has been built up and has proceeded. In this sense they are present in him, just as he is dependent on them. He is *imago mundi*. As "image of the world" he stands before God as the representative of all other creatures. He lives, speaks, and acts on their behalf. . . . Understood as *imago Dei*, human beings are God's proxy in the community of creation. They represent his glory and his will. They intercede for God before the community of creation. In this sense they are God's representatives on earth. (190)

Compare Moltmann's account of our double function as embodied yet capable of deliberate action to Descartes's position. Humans, Descartes argues, share with animals a "corporeal nature" characterized by instinctive impulses and a common "disposition of organs." But in addition to our animal part, we also possess a quality that was "expressly created," namely, our "incorporeal mind" or "soul," reflected in humans' unprecedented ability to use reason and communicate through speech. Furthermore, Descartes claims that animal physiology is formed in light of "special adaptation for every particular action," while humans are not confined to specific, created functions but have in reason a "universal instrument which can serve for all contingencies."[15] Humans, in other words, occupy the world at large, rather than remaining tied to particular modes of existence. All these points of Descartes's argument correspond to Moltmann's interpretation of humans' double function. The only departure from this basic scheme is Moltmann's greater emphasis on the dependence of humans upon animals.

The latter point regarding animals' spheres of adaptation deserves further consideration, for Moltmann's treatment of this idea is problematic as an "evolutionary" account. As we have seen, his concept of ecology of space resembles a pre-Darwinian economy of nature, assuming divinely bestowed fixity of created niches for animals. According to Moltmann, an organism's adaptation to its environment is a function of its originally created sphere of existence and the ongoing spirit of God in creation that maintains harmony in nature. This understanding of adaptation skips over the process of natural selection altogether. I have noted previously Stephen Jay Gould's argument that perfect design is not a compelling argument for evolution. Rather, the workings of evolution are best seen in the imperfect adaptations, the vestigial organs and ad hoc modifications. Moltmann's understanding of adaptation as perfect fit is not an argument for evolution by natural selection. If, as he maintains, adaptation to specific spaces provides each organism with a discrete range and function, then random variation, competition for finite resources, and differential survival rates within species would cease to be factors in evolutionary adaptation. It will be helpful to look at some non-Darwinian accounts of evolution in order to see how Moltmann's more closely resembles these than it does Darwinian theory. We will see that he is not sure how to treat natural selection, given his belief in the spirit as a guiding principle in nature that maintains community and adaptation. He posits suffering and struggle in nature as forces that the spirit of God overcomes, but it is not clear where these forces come from, since selection has no significant role to play in his theory of evolution.

"Adaptationist" theories of evolution such as Moltmann's have existed for centuries: Lamarckian theories of evolution, for instance, proposed a process of "orthogenesis" in evolution (literally, evolution in a straight line). According to this view, variation is directed toward the goal of adaptation. An organism's environment creates the animal's basic needs and these needs dictate how the animal uses its body. Through a process of use and disuse, an organism comes to fit its environment, and its acquired, adaptive traits are passed on to its descendants. Natural selection is rendered "powerless" in such a theory and "the species is carried automatically in the direction marked out by internal forces."[16]

In Moltmann's account this internal force is the spirit of God. One of

Darwin's contemporaries, botanist Asa Gray, proposed a similar sort of theistic evolution, arguing that God was responsible for evolutionary adaptation. Gray maintained that God has a hand in the forces that keep organisms adapted to their environments. His theory assumed, as Moltmann's does, that *any* natural mechanism that maintained adaptation was evidence of God's design.[17] Gray's process of evolution, like Lamarckian theories, would take a more direct route to adaptation, yet for the most part he accepted the general structure of Darwin's account of natural selection. He did not consistently deny the presence of natural, random variation, nor did he claim that no process of eliminating the unfit took place in nature. Rather Gray suggested that perhaps variation itself "directed" evolution, a suggestion rejected by Darwin, who saw that this interpretation would make natural selection redundant. Or, put differently, if selection continued, even in the presence of directed variation, any weeding out of the unfit seemed an unnecessary, wasteful step. Why should elimination of the unfit take place at all? Such a theory seemed to imply that the "Creator's hands [were] so tied that He could not design a more humane mechanism of evolution."[18]

Darwin's theory also revolves around a concept of adaptation, but evolution by natural selection involves *two* distinct steps: variation and "direction" in evolution are separate factors. First, there is random genetic variation—random in the sense that it is not oriented to an adaptive outcome. Natural selection then acts upon this random variation and those variants that are advantageous in the context of a particular environment are preserved. Variation is not oriented toward adaptation. The difficulty, theologically, for Darwinians is that if one associates God's activity with provision of the initial variation upon which selection acts, then the elimination of the unfit seems cruel and improvident. Doesn't God know which direction evolution will take? If one sees God's work in the "directing" force of natural selection, the problem still remains that widespread suffering is an integral part of the selection process. The attractive solution to such theological problems is to try to find some way of collapsing variation and selection into a single force that continually guides evolution toward adaptation.[19]

This, it seems, is what Moltmann's account of evolution has done: the spirit of God "creates new possibilities, and in these *anticipates the new designs and 'blueprints' for material and living organisms*," he writes. "In this sense the Spirit is the principle of evolution."[20] All created organisms "are directed towards their common future, because they are all, each in its own way, aligned toward their potentialities" (100). Moltmann's version of theistic evolution exhibits clear affinities to non-Darwinian theories. His uncertainty about how, if at all, natural selection affects evolution is apparent in his claim that God's indwelling spirit "drives out" struggle and strife (213). Natural selection becomes superfluous in this account. Although modern evolutionists may disagree on how significant a role natural selection plays compared to other factors such as genetic drift, all would agree that it plays some role. Yet there is essentially no role for it in Moltmann's account, other than its inexplicable presence as a minor chord that is continually brought back into harmony with the rest of nature by the spirit of God.

Not only does the spirit intercede in nature on a day to day basis, maintaining adaptive fit between organisms and their environments, but, over the course of creation as a whole, evolutionary forces that cause suffering are ultimately overcome. Recall Moltmann's account of the three stages of creation. Here evolution is synonymous with *creatio continua*—the intermediate stage between the world as originally created and creation in its ultimate, restored stage. Continuing creation occurs through the new possibilities, the new designs, that the spirit of God introduces in nature. Moltmann suggests that in the final stage of creation, *creatio nova*, all evolutionary processes that result in strife come to an end once and for all. Until the time of *creatio nova*, when the spirit "drives out the forces of the negative, and therefore also banishes fear and the struggle for existence from creation," *creatio continua* maintains as much harmony as possible in a fallen world. Ultimately, "the spirit who dwells in creation" transforms creation's "history of suffering into a history of hope" (102).

To sum up: Moltmann's account of evolution and the ecology of space allows us to underscore the following points. In attempting to replace a Cartesian, dominating, scientific perspective with a modern ecological perspective, Moltmann has failed on several counts. He has neither broken the spell of a Cartesian worldview nor offered in its place a viable alternative that draws from modern biological concepts and theories. Processes of evolution such as competition, predation, and extinction are virtually nonexistent in this account, aside from some vague references to suffering and struggle, which are themselves assumed to be only temporary conditions that will ultimately be banished. Moltmann's account assumes that adaptation generally prevails—at least, so long as humans do not destroy the order and harmony inherent in nature. Thus the role of natural selection is obviated and God's purposes can more readily be aligned with natural processes.

Before turning to process thought, one important question must first be raised regarding Moltmann's ecological theology. How did original creation become corrupted to the point of needing redemption and restoration at the time of *creatio nova?* If the spirit in creation dwells at all times in continuing creation (or evolution), promoting community and harmony, why is a new creation necessary? How did these negative elements insinuate themselves into creation in the first place? We have already seen that ecotheologians are often ambivalent about nature and its processes. As we saw with ecofeminists' arguments, their descriptions of nature leave out the presence of suffering and strife, even while their ethic aims largely at the removal of it. In fact, suffering and oppression are assumed to be so pronounced as to warrant wholesale liberation of nature. Ecotheologians are unsure of what constitutes nature's true nature.

Moltmann's work illustrates this ambivalence as well. He observes that Paul was correct in his account of the "whole enslaved creation," and he echoes Paul's invitation to Christians to recognize their "profound solidarity" with enslaved creation (67). Nature, like humanity, longs for liberty from these conditions. It too has "fallen victim to transience and death" because of sin. "A sadness lies over nature which is the expression of its tragic fate and its messianic yearning" (68).

Yet this account of fallen creation begs the question of why it fell and in what sense it is doomed to a "tragic fate." Process theology is better able to avoid this difficulty in its interpretation of God's role in evolution. But Moltmann's theology is less clear on this point. The implication is that creation was perfect in its original form (*originalis*) and became corrupted with the fall of humans. Were it not for this fall, nature would have remained harmonious; yet, even in its enslavement, the spirit works to foster whatever community and harmony remain possible until such time as creation, and all humanity as well, will be redeemed and liberated from transience and death. Moltmann's belief in a panentheistic God helps somewhat because it allows that "everything is not God; God is everything." In other words, God is present in all things, but not to the same extent. "God does not manifest himself to an equal degree in everything" (103). Does Moltmann believe that animals whose existence causes the most suffering, such as predators or parasites, have less of the spirit of God in them? Perhaps he believes that different degrees of divine immanence in nature are played out in harmonious and tragic elements.

In this discussion of nature as suffering and enslaved, what has happened to Moltmann's affirmation of nature as a community of harmonious interaction? Doesn't nature show a trend toward "universal symbiosis of all systems of life and matter" (205)? These ecological, community descriptions of nature appear to have more to do with a hoped for state of "liberty," when struggle and strife cease, than with nature's current condition (39). As in McFague's theology, we see that the ecological model as community is not so much a modern, scientifically derived perspective as it is an ideal account of what nature once was and will one day be again. Nature as it is currently (*creation continua*) only approximates the ideal ecological model. Suffering and discord in nature are not part of its original, ontological structure but emerge as something temporary and anomalous owing to nature's corruption. This feature of his ecotheology distinguishes it from process theology, which understands discord as an inherent feature of all life—and, especially, human life. We will now turn to a process view of nature and environmental ethics in the work of John Cobb and Charles Birch.

PROCESS THOUGHT AND THE ECOLOGICAL MODEL

Process thought refers to a broad and often abstract set of doctrines about the structure of reality, the nature of God, and the way in which God's will is manifested in the world we experience. We will begin with a summary of the basic features of process thought that inform the ecological theology of John Cobb and Charles Birch. There are many areas of overlap between process theology, ecofeminist thought, and an eschatological theology such as Moltmann's. For example, McFague's concept of God as a nonhierarchical and nontranscendent being who is put "at risk" when humans destroy nature owes much to process thought. Moltmann's theology also shares certain features with process thought, such as the hope for liberation of life and an understanding of creation as unfinished, although process thinkers do not necessarily emphasize Moltmann's messianic and trinitarian themes. Moltmann, ecofeminists, and process theology all draw on an "ecological model" of life as fundamentally interconnected and interdependent. The emphasis on relationality and the need to reject objectifying and mechanistic modes of understanding are central to all these schools of thought.

Although process theology did not emerge as a response to environmental degradation, many people have recognized its potential for promoting a better ethic toward nature, and Cobb and Birch have played a key role in explicitly developing these environmental themes. Process theology often begins with a rejection of creation ex nihilo, positing creation out of chaos, rather than nothing. The creation of order out of chaos is an ongoing process, not an event that occurred in the past. Since creation—or nature—has an essential unfinished quality, the future of the world is open to continually novel, created possibilities.[21]

Process theologians such as Birch and Cobb often draw on inferences from modern physics and chaos theory in support of their belief that there is a fundamental indeterminacy in nature. This focus can be traced to Whitehead who, incorporating findings from the new physics (quantum theory, relativity theory), proposed that the world consists in a network of interactions in which all events are interdependent. According to this view, absolute distinctions between subject and object are called into question and emphasis is placed upon all entities as *events* rather than substances. All things, living and nonliving, are constituted by their relationships. Birch and Cobb refer to Whitehead's perspective as an "ecological view of reality."[22] This understanding of nature is also endorsed by some contemporary biologists. Brian Goodwin, for instance, argues for a new kind of biology, a "science of qualities" and of "holistic emergent order," which "resonates with another myth, that of creation out of chaos."[23] But what happens to this science of qualities when it is translated into an environmental ethic? Should we value the qualities themselves? And if so, which qualities are most important? For his own part, Goodwin not only acknowledges capacities of subjecthood in individual organisms but also pays attention to species, ecosystems, and populations, regarding nature as a whole as a subject of sorts.[24] Citing Gunther Altner, Goodwin argues that "the prime obligation of human beings toward their fellow creatures does not derive from the existence of self-awareness, sensitivity to pain, or any special human achievement, but from the knowledge of the goodness of all creation, which communicates itself through the process of creation."[25] As I will argue, sufficient attention is not given to this larger picture of nature and its processes in Birch and Cobb's ethic; rather those individual qualities—self-awareness, capacities for pain, etc.—are abstracted from the larger processes from which they emerge.

Birch and Cobb's interpretation of reality asserts a basic unity and interaction to all life without sacrificing the importance and dignity of individuality. We have seen this understanding of the ecological model previously in McFague's account of the subject-subjects approach. Physical reality (and causality itself) is comprised of a complex *process* rather than an atomistic array of discrete objects, as the mechanical view held. In this sense process thought underscores continuity among all entities and undermines dualisms.

Rather than interpreting indeterminacy as proof of the meaninglessness or purposeless of the universe (as chaos theorists sometimes do), process thought embraces unfinished creation as the medium through which God interacts with nature, "luring" it—but not forcing it—in certain directions. God creates new possiblities for life: "It is God who, by confronting the world with unrealized opportunities, opens up a space for freedom and self-creativity."[26] In saying that God's will is manifested in nature, however, process theologians are wary of asserting that God's will is decisive: in the language of process theology, God acts persuasively, *not* coercively. As such, God's will and the will of humans may or may not coincide, but the possibility that they *will* coincide is always present. "If we decide to enter into the reality into which God calls us, we choose life. If we decide to refuse it, we choose death."[27] The very nature of God and creation makes hope possible; the same openness, however, makes fear equally possible and perhaps inescapable. In the context of current environmental degradation, this means that the prospect of humans utterly destroying their world is very real. Process thinkers hope for the eventual triumph of good without offering any guarantees of it. This uncertainty about outcomes contrasts with Moltmann's eschatology, which also alludes to the openness of creation (the spirit of God creates new possibilities) but is more certain that a renewed, peaceful creation is built into original creation. In other words, Moltmann's theology posits an eschatological triumph of good as a necessary outcome of the basic structure and progress of creation.

Process thinkers recognize certain affinities between their theological doctrines and evolutionary/ecological perspectives. In fact, Birch and Cobb are well aware that taking evolutionary theory seriously implies a different view of God and nature than traditional Christianity has assumed.[28] This awareness is reflected in their understanding of nature's interdependence, which, as we will see, is more complex and morally ambiguous than Moltmann's. However, even process theology quickly strays from its own Darwinian assumptions.

PROCESS INTERDEPENDENCE

Eschewing simple causation for an "infinitely complex" view of causality, process thought gives "primacy to interdependence as an ideal over independence"—interdependence in nature is "ontologically given."[29] There is a basic "ecological sensibility" to process thought, expressed in its awareness that every event within an ecological system is "made possible by a complex interconnection of antecedent events."[30] Ecology as the "study of the interrelationships of all things" affirms the process view of nature.[31] Likewise, the intricate workings of evolution can be understood in process terms because evolution operates in ways that produce new possibilities from complex antecedent events. Evolutionary developments create additional possibilities and ever increasing complexity. God, as the force that persuasively lures creation forward to new possibilities, confers some order to chaos and creates value within the context of evolutionary change: "Each stage of evolutionary process represents an increase in the divinely given possibilities for value that are actualized."[32]

Again, Birch and Cobb contrast this view of nature to the mechanistic account. The rejection of simple causality in favor of complex, ontological interdependence also entails a rejection of modes of thought that tend to compartmentalize life-forms, abstracting them from their surrounding environments. The importance of environmental influence on shaping all entities is repeatedly stressed in process thought and the ecological model to which it corresponds. The mechanistic model is valid only in cases where "structures studied are relatively independent of environment."[33] For most practical purposes, a stone, for instance, can be understood in the context of a mechanical model, even though it too is influenced by its environment. The separation of subject and object that pervades our worldview (or, at least, Western and scientific worldviews) is shown to be false. The emphasis on interdependence and interrelationship does not, however, preclude understanding suffering as an inherent part of life; this insight brings process thought closer to a Darwinian perspective than other strands of ecotheology we have looked at. Let us see how process theology explains the presence of suffering and evil in the world.

PROCESS VIEWS OF SUFFERING, EVIL, AND EVOLUTION

Suffering and evil are explained in terms of human freedom and the interplay between what process thinkers refer to as "enjoyment" and "discord." In our evolutionary history increasing complexity—which is good in itself and creates the possibility of enjoyment—also brought about the possibility for evil. Process thought asserts that evil is not necessary, but the *possibility* of evil is.

As evolution proceeds, consciousness acquires greater and greater complexity. "The emergence of conscious experience was the crossing of a great new threshold," Birch and Cobb argue, "which we associate with the development of a central nervous system."[34] Thus, organisms with central nervous systems can be assumed to possess greater richness of experience than other life-forms. Capacity for enjoyment also increases with greater conscious experience. All animals direct their conscious experience toward the goal of meeting their bodily needs, but some higher animals also possess a great capacity for "self-enjoyment" even while the primary function of their experience is "service to the body."[35]

Humans are different in this respect: very little of our activity is carried out with specific regard to benefits for our bodies. Furthermore, the human brain reached a second threshold, beyond that achieved by animal consciousness, when "unified human experience began to make its own enjoyment its primary end, and to instruct the body accordingly."[36] Once this stage of evolution was reached, our bodies were no longer restricted in their function merely by external forces (as animals are restricted by their environments.) Now, however, we are restricted just as powerfully from within, by our own psyches.[37] Humans, in other words, have a psychic existence unknown to animals and therefore have a more complex dimension to their existence. Birch and Cobb associate this second threshold in evolution with a human "Fall," which introduced, for the first time, widespread *dis*harmony in the natural world but also made a richer life possible. This interpretation of the Fall has the merit of at least acknowledging that humans have not existed from the time the earth was created.

In our evolution increased consciousness and psychic complexity entailed the possibility of deviation from good as well as richer experience. Since God is the ground for all possibilities, God is in a sense responsible but not indictable for evil in the world. Process thought "distinguishes between divine responsibility and blameworthiness."[38] In human beings, who among animals possess the greatest degree of complexity, discord (defined in terms of physical or mental suffering) and enjoyment go hand in hand. If we choose to live so as to minimize discord, we stand to lose much that also makes life enjoyable and risk being left only with "triviality"—the process term for existence characterized by the absence of pain without any significant enjoyment. One may opt for as much harmony as possible, averting discord, but this involves a reversion to a more trivial kind of existence, devoid of the intensity of experiences that enrich life. Intense enjoyment is impossible without some degree of suffering and discord. Process thought does not deny that suffering is real; it simply explains its existence in terms that are consistent with the existence of human freedom, the evolution toward complexity, and the persuasive rather than coercive will of God.

The basic account of the positive benefits of struggle and strife does not apply only to human beings, even though humans have higher-grade experience. Birch and Cobb suggest that the existence of strife is met with the possibility of transcending it in all life-forms (even though, presumably, discord in animals pertains more to the body than the psyche.) That is, Birch and Cobb do not deny that animals have some richness of experience but distinguish this experience, which is geared primarily toward attaining bodily harmony, from human experience. They suggest that evolution by natural selection operates by provoking in animals a transcending response in the face of environmental pressures that produce discord. "If this transcending were wholly absent from the rest of nature, it would introduce such a fundamental cleavage between human beings and other living things. . . . But according to the ecological model, all living things share in this experience of transcending."[39] Conditions of discord in nature generate "new responses" in all life forms which, in turn, create novelty: "Living things inherit from their physical environment, but their response introduces something new."[40] In this way novelty is "introduced in the evolutionary process."[41] Again, process theology does not *guarantee* that suffering will ever be transcended completely. Yet, the removal of suffering—not just "unnecessary" suffering—remains a central value in much of Birch and Cobb's writing, despite their claim that "contentedness and discontentedness belong together."[42]

Important questions are raised by process theology's account of evolution and its insistence on the inextricability of suffering and enjoyment. First and foremost: why is *liberation from suffering* held out as the ultimate hope for all life? The ecological ethic Birch and Cobb formulate revolves around reducing the suffering of beings who possess richness of experience. Given its sophisticated account of the existence of suffering as a corollary of enjoyment, it is not always clear why a process thinker should endorse this ethic. Suffering is "necessarily entailed in the creation of beings capable of high grades of enjoyment."[43] Thus, God does not banish all suffering, as in Moltmann's account, but divine love "encourages us to avoid *unnecessary discord*."[44] Once explained in process terms, it would seem that suffering ceases to be a "problem" at all. After all, it is in part the process of "striving and struggling" that makes life worth living—"what is needed is not a tensionless state" (108).

The relationship between the goal of liberation and other features of process thought becomes especially problematic in some of Birch's writings in which he suggests that liberation *is* in fact the guaranteed outcome, not merely a hope for creation. As we will see, Birch invokes a biblical ideal of nature without suffering, a vision of the natural world restored to a pre-Fall "paradise." It is difficult to reconcile the process account of evolution toward complexity (and hence, evolution toward both enjoyment and discord) with a hope for a restored paradise. Other difficulties with process thought emerge when viewed from the standpoint of evolutionary theory. For instance, what is the relationship between the liberation paradigm and the evolutionary understanding of life that Cobb and Birch profess to embrace? How is evolution understood vis-à-vis the "ecological model" that is central to Birch and Cobb's conception of life? Do *evolution* and *ecology* refer to the same view of nature?

We will turn to this last question first in the analysis of process theology that follows, paying close attention to Birch and Cobb's use of the terms *evolutionary* and *ecological* as well as their definition of the term *liberation*. We will see that ecology and liberation are complementary concepts (this argument has recently been made even more forcefully by McFague, who maintains that liberation theology and ecological theology are one and the same.) It is less clear, however, where evolutionary theory fits into the process account of the ecological model and the goal of liberating life. Following this discussion, we will look more closely at Birch and Cobb's account of richness of experience and their use of this concept as a basis for an environmental ethic.

EVOLUTIONARY THEORY AND THE ECOLOGICAL MODEL

The ecological model signals in process theology (as it does for McFague, Ruether, and Moltmann) an understanding of nature as interdependent and interrelated. However, the process idea of interdependence involves an explicit acknowledgment of nature's dynamic and complex causality: interdependence does not necessarily imply a simple or static harmony in nature. The evolution of conscious experience plays a central role in the process account of suffering and discord and, as I have previously suggested, their incorporation of suffering as an integral part of nature's interdependence distinguishes Birch and Cobb's approach from other ecotheologies we have considered.

Birch and Cobb appear more wary than most ecotheologians of lapsing into romantic paeans about nature's balance and order. At the very outset of their collaborative work *The Liberation of Life*, they address the significance of evolutionary theory: "We take evolutionary history seriously in the effort to understand humanity, its future, its relations to other living things and to the life that is inwardly experienced" (3). A close relationship exists between evolution and ecology, they argue; indeed, evolution should be understood "ecologically through and through" (4). They argue that both evolution and ecology draw on a central concept, often overlooked by environmentalists, namely, "the struggle for existence." Moreover, they recognize that struggle in nature is apparent on many levels—between individuals, between an individual and its environment, among different species. Struggles that involve suffering of individuals, such as those between predator and prey, are perhaps the most obvious example, but in nature everything that affects the existence of an organism involves struggle. Birch and Cobb are refreshingly critical of environmentalists who blithely celebrate the web of life and the balance of nature. Environmentalists need to use "more precise language" and must recognize that Darwin's ideas seriously challenged the pre-Darwinian concept of balance in nature. Evolutionary theory, they observe, has

> had considerably less influence on ecological theorists of the "balance of nature" school than they should have had. . . . Sound ecology will depend less upon glib phrases such as "balance of nature" and more upon real observations and experimental manipulations in the field. (41)

Darwin himself drew on a "web of nature" concept in explaining his theory, they note, but, in his view, part of what joined life in this web was the common strands of struggle. Relationality should not be taken to mean harmonious and cooperative relations. Thus for Birch and Cobb, far more so than McFague and Ruether and Moltmann, *interdependence* emerges from complex relationships, including those of conflict and struggle for existence; it is not simply depicted as the counterbalance to conflict that keeps it at bay. The struggle for existence is the raw material from which nature's interdependence is constructed insofar as struggle produces new "responses" in its wake, such as individual adaptation to the environment. These responses, in turn, reinforce interdependence. Birch and Cobb recognize that evolutionary and ecological interdependence does not present us with a "stable, harmonious nature to whose wisdom humanity should simply submit" (65). On the whole, Birch and Cobb's account of evolution, struggle, and interdependence is vastly superior to the picture of nature that we find in some other ecotheologies. They warn against the tendency to separate concepts of evolution and ecology, and, when speaking of the ecological model, they intend that evolution must retain a central role.

But the ecological model has broader implications than the evolutionary concepts that are assumed to make up its core, for this model contains within it the very principles needed for liberating life—and the life sciences. An ecological model "pictures the organism as inseparably interconnected with its environment" (80). Consistent with tenets of process theology, this model holds that the environment of an organism is *constitutive of* that organism. If, as Birch and Cobb argue, ecology is the science of relationships, then it is clear why process theology, which insists that relations are fundamentally constitutive of entities, necessarily embraces an ecological model. It is less obvious, however, why process theology or the evolutionary perspective as they interpret it, should promote an ethic of liberation.

LIBERATION OF LIFE AND THE LIFE SCIENCES

For these authors process thought, ecology, and liberation are inseparable concepts. Birch and Cobb uphold the ecological model as the better alternative to the objectifying, categorizing approach of the mechanistic approach—what McFague calls the "arrogant eye" of Western science that disregards animals as experiencing subjects. The ecological model is a model *of life* and, as such, it is a liberating model: it aids in the liberation of the very *concept* of life from oppression and objectification. Conceptual liberation is the first step to liberating life itself. Birch and Cobb thus distinguish conceptual liberation (liberating the way we conceive of life) from more direct, active forms of liberation (the eradication of social structures and human behaviors that perpetuate oppression and suffering). The ecological model, understood in the first sense, liberates the discipline of science from its own ancient oppressive roots.

When we reflect upon some of Birch and Cobb's characterizations of the ecological model, it appears, once again, that the model is both derived from nature and extended to it as a means of liberating life from oppression. The impression that the ecological model is extended to nature from some other source is reinforced by Birch and Cobb's claim that ecological interrelationship is most clearly seen in *human*-human interactions: "Our proposal is that something important can be learned about all living things by taking human experience as the prime illustration of the ecological model" (105). In fact, they assert that the important question "is not whether the ecological model applies to human experience but whether, in its fully expressed form, it applies to other animals" (122). This seems an odd question to pose, given the term *ecological* model.

In any case, Birch and Cobb argue that this model can help to liberate scientific thought itself, creating different conceptual foundations for science, and that it can help to liberate all life in more direct ways. The ecological model, in other words, supports an ethic of liberation, not merely a new way of thinking about life. In what does this ethic consist and to which kinds of beings does it apply? Answering this question involves a closer examination of Birch and Cobb's criterion of richness of experience for assessing the value of all organisms.

RICHNESS OF EXPERIENCE AND ENVIRONMENTAL ETHICS

"The ecological model," Birch and Cobb argue, "applies to human beings as well as to all other things" (139). As a basis for understanding all life, it entails the "liberation of social structures and human behaviour such as will involve a shift from manipulation and management of living creatures, human and non-human alike, to respect for life in its fullness" (2). The liberation of life from oppressive structures must occur at all levels of life, from the individual cell up to and including human communities. Given current scientific perspectives, they propose that life exists on three basic levels: the cell, the individual organism, and the population level, arguing that liberation "is called for at every level of biology" (ibid.). But what constitutes "oppression"? Birch and Cobb acknowledge that liberation is often defined simply in circular fashion, that is, in terms of what it wants to get rid of, namely, oppression. They concede that there are obvious limitations to defining "liberation" and "oppression" in this way. Still, their "faith in life" justifies their belief that in the absence of oppression "something very positive happens," even if they are not always able to define what this is: "Clarification of what life is and what it means to trust life will support the confidence liberators long for" (3).

Despite the authors' recognition that "liberation" and "oppression" are too often defined simply with reference to one another, they do little to develop these concepts beyond their circularity. It is not clear in what sense life *as a whole* is in need of liberation from oppression. In fact, the vast majority of *The Liberation of Life* deals with human communities and their oppression of one another. The predominance of chapter headings devoted to human ethics and societies (such as "The Biological Manipulation of Human Life," "Economic Development in Ecological Perspective," and "Rural and Urban Development in an Ecological Perspective") illustrates that Birch and Cobb's primary concern lies in applying what they understand as the ecological model to human life. This focus is consistent with their claim that the model is best exemplified by human relationships, but it takes them further and further afield from their ethical interest in life "as a whole."

What qualities do human and nonhuman life-forms possess that unite them and make all life deserving of liberation? Conscious experience implies greater capacity to suffer. Liberation entails removing or reducing suffering in beings who have high degrees of richness. Or, as Birch and Cobb sometimes express it, beings with high-grade experience have more rights than those without it, and our ethical obligations are greater toward these. Thus, liberation, richness of experience, capacity for suffering, and possession of rights are connected. We have seen that Birch and Cobb maintain a basic continuum of richness of experience between humans and animals and that they envision evolutionary process, or perhaps progress, as the driving force behind increasing richness.

The ecological model teaches that humans are subjects in a "wider community of subjects" and that there exists "a continuity between all levels of existence" (151–52). Our radical interdependence with all life implies that we have a "moral responsibility, beyond the scope of the human community." Our responsibility toward other forms of life is connected to the claim that all life consists—in varying degrees—of experience. Those beings that have the greatest richness of experience are "more alive" (146), and we should respect and show concern for animals "in proportion to their capacity for rich experience" (153). Thus Birch and Cobb propose a "hierarchy of value" based on richness of experience "from the simplest forms of life through to human beings corresponding to the richness of experience of each form of life" (205). The ethical requirement that accompanies this interpretation of value is that we are to "promote" richness in all life. However, the authors concede that since it is difficult positively to *produce* richness of experience in nonhuman animals, we should instead concentrate on *not inhibiting* richness of experience. Since a large part of animals' experience—even though rich—is oriented toward bodily enjoyment and harmony with external environments, our obligations toward them consist in "reduc[ing] the amount of pain and suffering we inflict upon other creatures" (155). A liberated and liberating ethic of life, therefore, leads to an ethic of reduced animal suffering.

Animals that have the greatest richness of experience (those with sophisticated nervous systems, larger brains, complex mental and emotional lives) are those toward whom our moral obligations are the greatest. Such organisms cannot be viewed as means only but are also ends in themselves. Birch and Cobb propose that all life, including human life, can be valued as both ends and means, having varying degrees of intrinsic and instrumental value. Intrinsic value is assessed according to richness of experience; therefore some organisms have great intrinsic value while others have a negligible amount. Those that have little intrinsic value "may reasonably be treated as means, or in terms of their instrumental value only" (152). Plants, for instance, have great instrumental value for those things that depend upon them but little intrinsic value: "We judge that plants, like the cells that compose them, can appropriately be treated primarily as means" (153).

At this point, however, Birch and Cobb's argument has already traveled a considerable distance from the ecological, evolutionary assumptions that were to have informed the ecological model. This is apparent when we reflect upon the following points, which, before turning to a detailed discussion of them, I will summarize here. First, it is difficult to reconcile conferral of value along lines of richness with an evolutionary/ecological understanding of nature. The ethic proposed here gives more value to organisms with richer experience. An ecological, evolutionary perspective would not assign value to animals according to their status as "experiencers." Organisms—and groups of organisms—play important, valuable roles in ecosystems and as predators and prey of other species, regardless of whether or not they have a rich subjective life individually. Birch and Cobb provide no adequate grounds for valuing organisms with little or no richness.

Second, the centrality of the idea of interdependence is violated, in two ways, by this interpretation of value and the ethic it generates. As already suggested, the richness criterion could lead to preferential preservation of those species that have high-grade experience. The concept of interdependence demands that continuity and dependence of all living things be preserved and respected. Yet richness of experience draws lines between types of organisms based on a human-centered criterion of value (much, it seems, as the old mechanical model of nature would have done). Furthermore, if interdependence is built upon, and emerges from, *all* relationships, including those of conflict and strife, and if strife produces important, transcending responses in organisms, then it is not clear why our ethic should consist in *reducing* their suffering. Birch and Cobb are not alone in making this dubious connection between the presumed structure of nature and the ethic of liberation (as I will stress in a moment, Rasmussen makes comparable claims).

Finally, regarding Birch and Cobb's account of evolution as a process of transcending responses, there are reasons to believe that this interpretation of evolution is not compatible with Darwinian theory. It is, rather, an affirmation of the importance of the environment as the crucial, influencing factor in eliciting creative responses from individual organisms, as described by the ecological model. This interpretation of the environment as *directly* molding organisms who respond to changing, external factors is not necessarily consistent with evolution by natural selection, as we will see.

RICHNESS VERSUS BIODIVERSITY

Essentially, Birch and Cobb's ethic—not unlike McFague's—values animals in proportion to their likeness to humans. The difference is that Birch and Cobb explicitly acknowledge that they accept a hierarchy of values in animal life, whereas McFague's ethic unwittingly results in a ranking of subjects. At certain points in Birch and Cobb's argument their bias in favor of humanlike qualities is quite clear: they argue, for example, that a much more serious transgression is committed by killing a chimp than by killing a chicken. Among the former, "there are indications of an individuality resembling our own and of social relations which lead to grieving for the dead" (160). "Rights" for animals are assigned according to such characteristics (154). Yet Birch and Cobb maintain that the ethic they propose entails a "rejection of anthropocentrism," despite the fact that human capacities and experiences provide the reference point.[45]

However, there is no reason to suppose that organisms richest in experience are the ones whose representation in nature is most important for the preservation of an ecosystem. In other words, the richness of experience criterion cannot function as a *biodiversity* criterion. This ethic, if applied to nature, would give priority to species who possess greater consciousness. "To us," they write, "it is much worse to inflict suffering on creatures with highly developed capacity to suffer than on those where this capacity is rudimentary."[46] Note that Birch and Cobb often speak of our obligation to avoid *inflicting* pain. This might mean that we should reduce the suffering of animals in nonwild situations—situations where humans are directly responsible for "unnecessary" suffering, such as unjustified animal experimentation, mistreatment of "food" animals, animals used in entertainment, and so on. But their argument demands more than this. They speak of obligations toward all life. Like McFague's, this understanding of life is all-inclusive, involving, as we have seen, the liberation of life at every level and every place, from the cell to the community. Thus it seems that the ethic of reduced suffering is not meant to apply merely to anthropogenic sources of suffering or the suffering of nonwild animals, but also applies to the suffering and rights of animals in nature.[47]

Recall, for instance, Birch and Cobb's claim that greater rights attach to animals with the highest-grade experience. To show how relative claims are assessed, they cite an example involving degrees of value in different forms of marine life. Porpoises exhibit all the signs of having a "rich subjective life" and therefore "we can confidently affirm that the intrinsic value of a porpoise is greater than that of a member of these other species [tuna and sharks]."[48] Accordingly, they conclude, porpoises have greater rights and our obligations are greater toward them. If we must choose between enhancing or preserving the life of porpoises versus preserving the life of tunas, we ought to choose the porpoise. Such an ethic is utterly inappropriate in an ecological, evolutionary context. As Holmes Rolston argues, humans have no duty to "promote by intervention the richest psychological life of wild animals, or their richest biological life." If we assume such duties toward nonhuman life, would this mean "adjusting the number of rabbits that the coyotes devour? Should we divide the harem of a dominant bull for the more equitable satisfaction of young bulls, promoting the maximum welfare of each male?"[49]

The difficulties of reconciling a Darwinian account of nature with an ethic based upon richness is especially evident in some of Birch's work distinct from his collaborations with Cobb. Here Birch has explored the theme of liberation from oppression in greater detail, arguing, despite the focus in Christianity on concern for oppressed classes of people, that Christians have generally overlooked that nature, too, suffers from oppression. "Nonhuman animals," including animals used in experiments and those in the wild, constitute "an oppressed group."[50] Birch draws on the same richness criterion in making his case that all animals should be regarded as ends as well as means. Their intrinsic value is a function of their capacity for experience, which, in turn, "may well be related to the complexity of the nervous system" (59).

Birch clearly has all animal life in mind, not just domestic animals or pets. He notes that many people think of their cats and dogs as capable of joy and suffering but tend to "draw the line" there, not realizing that "wherever we find a nervous system we may suppose there is something akin to what we call feeling." This suggests that "a wide range of vertebrates may experience some sort of suffering akin to anxiety in humans" (59–60). Because so much of human activity destroys life, the ethic before us is one of restoration and salvation. "Salvation," he writes, "is an ecological word because it is about restoring a right relationship that has been corrupted" (61).

Birch's turn to the language of *restoration* and *salvation* of nature signals a closer link in his own work between process thought and an expectation of eschatological eradication of suffering. Without Cobb's more sophisticated theology to guide him, Birch's ecotheology is less coherent *and* less accurate scientifically. That is, his environmental argument here rests on an assumption that suffering in nature is symptomatic of nature's "fallenness" and is therefore a temporary condition of life that will be corrected at the eschaton. Although process theology in general eschews assurances of the ultimate triumph of good over evil in the world and understands suffering and discord as built into nature, Birch's arguments display a greater degree of certainty that suffering can and ought to be removed. As we will see, his argument implies a disruption of natural processes in order to prevent suffering in nature—not just the suffering that human activity directly causes.

Like McFague, Birch proposes that Christians ought to understand their obligation to nature in terms of an ethical orientation that is uniquely Christian: nonhuman animals are our neighbors because a "Christian biocentric ethic takes the neighbor to be all that participates in life" (63). Indeed, there is an evolutionary basis for extending the neighbor ethic toward nature, in that an "evolutionary perspective leads to the concept of a continuity between all levels of life" (66). In keeping with the ethical implications of evolution and the Christian duty to love, we should "seek to be neighbor to nonhumans in a way analogous to the way we seek to be neighbor to our human fellow creatures." Also, like McFague's extension of a love ethic to nature, Birch's account of obligation draws on the healing ministry of Jesus. Christians are to "succor those who fall by the wayside and to try to remove the causes of suffering and to provide a room in the inn" (65).

There is no obvious connection between Birch's account of evolutionary considerations and his ultimate aim of removing suffering and healing organisms.[51] His ethic implies that he perceives the very processes of evolution to be products of a postlapsarian world. The Christian who practices neighbor love toward nature participates in the reduction of human-inflicted suffering. But, more than this, Christians are participants in an ultimate eschatological (re)establishment of the peaceable kingdom. Birch's claims here suggest that his concern is with eradication of all suffering in nature, natural or otherwise. For Christians who heal animal suffering

> the day is foreseen when paradise is regained, and everyone not only goes back to a nonmeat diet, but the friendliest relations subsist between all species. The wolf shall dwell with the lamb, the leopard shall lie down with the kid, and a little child shall lead them![52]

This ideal account of animal-animal and human-animal relations is a particularly odd one for a biologist to articulate. As the basis for an ethic toward nature, it is doomed to fail because it seeks to "return" nature to a state in which evolution itself has been eradicated. Stranger still is his suggestion that all animals were vegetarian prior to humans' introduction of evil into the world.

Birch fails to distinguish conceptually 1. the suffering entailed in practices such as meat production, animal experimentation, and zoo keeping and 2. the suffering that is inherent in nature. Put differently, he does not attempt to formulate a different ethic for animals in the wild and animals in other contexts. "The guiding principle proposed is that we are morally obliged to reduce suffering and to enhance the quality of life of animals that share the earth with us," Birch argues, and "greater obligation is entailed toward those creatures that have more significant experience."[53] The broad category of animals "that share the earth with us" is surely meant to include animals in nature.

Consider for a moment what would happen if we followed this ethic *in practice*. Let us assume that we have the greatest ethical obligation toward animals with significant richness of experience, such as those with complex nervous systems. The inordinate public concern with preserving the most sophisticated and complex animal species perpetuates endangerment of other, less "interesting" species. Environmentalists often have a difficult time getting people to care about—and donate money to—preserving species that are not particularly cute or interesting to behold.

For example, campaigners for environmental groups in the UK recently underscored the need to channel more money into preservation of invertebrate species, "the unsung heroes of the natural world."[54] These "lowly 'minibeasts'" attract little attention, yet their continued existence is absolutely vital for the survival of wildlife as a whole. Conservationists note that a number of invertebrate species are in serious peril, including various species of beetles and ants, because of the greater interest in preserving endangered, higher-profile vertebrate species. Estimates indicate that more than 150 invertebrate species have become extinct in the UK in the past 200 years.[55] An ethic such as Birch and Cobb's, which places much greater value on complex species than on invertebrates and plants, would be powerless to redirect concern toward "lower" organisms that have little or no intrinsic value.[56]

The extinction of invertebrate species other than insects has also generated little public concern recently, even among those who might consider themselves environmentalists. Freshwater mussels, for instance, are currently one of the most endangered animals in the United States, "the victims of our national penchant for damming, dredging, building on, and otherwise degrading our rivers and streams in the name of development or recreation."[57] Currently 10 percent of freshwater mussel species are extinct and 69 percent are protected under the Endangered Species Act. But most people have shown little interest in mussels. The general public does not easily "get worked up about a cold-blooded, gluey morsel of mollusk flesh" with "brains of only a few neurons."[58] Yet the extinction of "unlovable" animals affects the entire river ecosystem.

What would Birch and Cobb say about this case? Even if they grant that mussels might be preserved because of their instrumental value (which seems doubtful, since they spell out no moral obligations toward organisms without significant experience), their understanding of human richness and capacities for enjoyment implies that our own interests (in recreation, development, dams) would prevail. The enjoyment and enrichment of life that such practices afford human beings would seem greatly to outweigh the value of a mussel species. Furthermore, as we have seen, Birch and Cobb maintain that organisms with greater richness have greater "rights." If mussels have any rights, these are surely eclipsed by the rights of humans, who have the highest grade experience of any organism. If a porpoise has greater claims to rights than a shark, then surely human rights are more important than those of a mussel with little or no experience. Human responsibility "is only to the other experiencers"; we have no responsibility toward "things that are supposed to be devoid of the capacity for enjoyment."[59]

In order to see what is wrong with the process view of value, imagine a place devoid not only of human beings but also of major, complex flora and fauna, where even the idea of an ecosystem with hierarchical structures of different kinds of organisms makes little sense, where many of the native land species are mites, lice, and other small and generally parasitic creatures, and the rest are microscopic. If such a place existed, what sort of environmental ethic would make sense there? Even an ethic that is nonanthropocentric, one that takes intrinsic value of all organisms seriously, would become strained under such extreme circumstances. Yet Antarctica is just such a place.[60] And while it is an unusual place ecologically—perhaps a peripheral case ethically—the extremity of it forces environmentalists to think hard about what it means to value nature when it bears little resemblance to, and confers few obvious benefits for, human life. Where life is driven to extremes, so is environmental ethics: "Life pushing into those extremes does deserve our respect when we encounter it, and demands more vigilance, lest we disturb it."[61] Yet an ethic based on evolutionary complexity and richness as Birch and Cobb define it might easily dismiss such a place—an entire *continent*—as having little or no significant value.

We can also imagine numerous scenarios in which conflicts might arise between humans and animal species, as when reintroduction of predators puts human lives at risk. In all such cases Birch and Cobb's guiding principle indicates that human interests *always* take precedence. This point is underscored in an example cited by Birch and Cobb: Ruandans [*sic*] face a situation in which they must choose between "having elephants in national parks and land that can be farmed for people."[62] Without the use of this land for crops, the growing population of Rwandans faces an increased risk of starvation, suffering, and death. The needs (or "rights," as they might say) of the elephants and the human inhabitants are in conflict. Or, to characterize this scenario somewhat differently, let us assume that there are actually three kinds of entities involved—the elephants, the humans, and the land (which is an aggregate of things with only instrumental value such as plants, trees, and rocks, etc.). Two *experiencing* species are competing for access to something with instrumental value.

How is the conflict to be resolved? The land itself has great instrumental value—both to the elephants and the humans—but humans, with greater rights than the elephants by virtue of their greater richness, have claims that are more significant. If serious food shortages occur and the population continues to increase, "then in time we would reluctantly judge that Rwanda would be morally obligated to abandon its protection of elephants."[63] This outcome, however troubling to Birch and Cobb, is hardly surprising given their criteria for what constitutes value in life. An elephant may have immensely rich experience, but, in the hierarchy of value, it still ranks below humans. Birch and Cobb hasten to add it is imperative that humans restrain their growth in order to minimize such conflicts, and they frequently call for compromises between human interests with those of other organisms. Maximizing richness puts emphasis "on the quality of human life," they admit, "but not without serious consideration of the cost to other life."[64] But however seriously other organisms' interests might be considered and however "reluctantly" Birch and Cobb might agree with the decision to abandon the preservation of elephants, it appears to be a foregone conclusion.[65]

As we have seen, it is not clear how the basic ecological and evolutionary principle of interdependence as defined by these authors supports the ethical directives they endorse. Their conferral of value to organisms is abstracted from an ecological context and their concern with suffering in even wild animals (more pronounced in Birch's work) is also at odds with their own, clearly articulated, Darwinian account of interdependence. Recall the premise with which the authors began, namely, an understanding of natural interdependence as woven from threads of struggle, suffering, and conflict. We might infer from this perspective that suffering should be interpreted as crucial to life in the ecological model. The very fabric of nature's interdependence would fall apart if interactions that entail suffering in nature were impeded.

Furthermore, Birch and Cobb's account of the evolution of conscious experience, which involved the inextricability of suffering with enjoyment, implies that humans and at least some animals cannot have one without the other. Why, then, should we liberate life from suffering? Are we to deny animals the very transcending experiences that propel evolution of consciousness forward? As Cobb argues:

> We see that the development of beings with the capacity to enjoy significant values, and to contribute significant values to those beyond themselves, necessarily meant the development of beings with the capacity to undergo significant suffering, and to contribute significantly to the suffering of others beyond themselves. The good cannot be had without the possibility of the bad. To escape triviality necessarily means to risk discord.[66]

There is an "inextricable mixture of evil with good in all of life."[67] One wonders how much suffering can be removed without compromising what is good about life. Perhaps we ought to encourage suffering if it challenges organisms and evokes transformative, transcending creative responses.

Even the most scientifically minded ecotheologians sometimes make the mistake of uniting a sound evolutionary perspective with an ethic that essentially calls for an end to the very factors and experiences that are integral to evolutionary processes. An example is Larry Rasmussen, whose *Earth Community, Earth Ethics* shows a remarkable grasp of contemporary debates about ecosystems, concepts of balance and stability, chaos and disequilibrium. He notes that "just when 'ecosystem' becomes a common notion that makes the evening news, supermarket ads, and even books in religious ethics, some scientists quit using it."[68] This has serious implications for ecological ethics. Rasmussen points out, for instance, that the often touted goal of sustainability is in reality difficult to define, much less apply, given what we now know about ecological systems. Increasing evidence of the fundamentally erratic, unpredictable, fluctuating nature of ecosystems suggests that the very concept of sustainability is far more complex than we realized. The idea of sustaining, and even concepts such as "scarcity," imply that there is a background order or stability, a baseline, at which environmental stewardship aims; such ideas emerged from outdated, static, equilibrium-oriented models. "Uncertain science chasing uncertain nature" is what we have to contend with in contemporary ecology (167).

His discussions of evolutionary theory are also well-informed. Rasmussen recognizes the "moral ambiguity" of nature, as well as the ambiguity that inheres in our ethical choices regarding the natural world (247). "None of the choices is made easier," he writes, "by science's portrayal of the strange unity of life and death that comprises creation's integrity. Earth is the source of all life, yet the abode of death—and both indivisibly" (246). Rasmussen's recognition of the inseparability of creative and destructive forces in nature echoes the process view that harmony and discord, suffering and enjoyment, are two sides of the same coin. Moreover, he is careful to point out that it is not only *human* beings whose survival and way of life entails encroaching upon, even exploiting, other forms of life. He avoids Ruether's romanticized version of nature as observing limits, while humans deviate from natural norms, introduce destructive forms of competition, thwart natural limits, and dominate other life-forms. Under *normal conditions*, he writes, life (and not just human life) "takes advantage of life, exploits the environment, treats neighbors unfairly, and then leaves it to them to handle the waste in recurring cycles!" (ibid.). He articulates an evolutionary cosmology that has come to terms with suffering, disorder, death, as built into—rather than aberrations of—nature's interdependence and relationality. Some forms of suffering make life difficult, he notes, but "we cannot *live* without it. It belongs to our creaturely constitution to struggle in the process of becoming" (289).

But where he goes astray is where Birch and Cobb (of whom he approves) also go astray, namely, in joining an evolutionary cosmology (for the most part, accurately described)—a cosmology explicitly premised on the inseparability of life and suffering—with an ethic that seeks to overcome suffering and death, to "redeem" and "liberate" life. "In my judgment," he writes, "an evolutionary sacramentalist cosmology offers the richest conceptual resources for addressing earth's distress, if infused with a profound earth asceticism and married to prophetic efforts aimed at 'the liberation of life: from the cell to the community'" (247). This last phrase is taken from the title of Birch and Cobb's collaborative work. Despite the recognition that life under normal conditions is bound up with struggle and death, Rasmussen calls for redemption of "all creation," a "radical transformation of the created order," and deliverance from "whatever oppresses or victimizes" (256). Redemption involves the gift of life and its blessings to all creatures, an "ecumenical and ecological ethic of life," in pursuit of "healing, mending, and transforming the world" (257).

This redemptive, healing agenda seems at cross-purposes with an evolutionary perspective that acknowledges and, in some sense, even celebrates the inseparability of life-giving and life-taking forces in the natural world. The cross of Jesus stands in opposition to "most" but "not all" suffering, Rasmussen writes. Like Birch and Cobb's delineation of "*unnecessary* suffering," Rasmussen too suggests that some suffering is "negative" while other kinds are indispensable "in the service of life itself" (289). Which is which, and how do we know? At times, Rasmussen, like McFague, implies that we are obligated to help and to heal whenever and wherever the need arises, both in human and natural communities. Also like McFague, he contends that organisms can be viewed as our "neighbors," no less deserving of justice and love than their human counterparts. "Neighbor is as neighbor-love does to whomever or whatever is at hand and in need. . . . The neighbor in a million guises is the articulated form of creation to whom justice, as the fullest possible flourishing of creation, is due" (261). How much "flourishing" is realistically possible? How much suffering is inevitable? These statements need greater clarification, in light of Rasmussen's otherwise commendable attention to evolutionary and ecological realities.

On what basis, then, can ecotheologians maintain that the life-promoting, liberating ethic they propose is consistent with modern biological concepts? Sometimes, as is the case with Birch and Cobb, they have a very specific, and highly questionable, interpretation of evolution in mind. Their primary interest in evolution lies not in the Darwinian account of interdependence that they articulated at the outset but rather in the process understanding of the evolution of consciousness that they adapt from Whitehead. This interpretation of evolution is not necessarily evolution *by natural selection*.

THE EVOLUTION OF CONSCIOUSNESS

Process theology values increasing complexity in all life-forms. Increased complexity, the evolution of consciousness, and greater richness of experience express the creative side of nature as well as the persuasive, luring action of God. The unfulfilled possibility of all life is, at every moment, the basis for openness—and therefore hope—for the future.

Birch and Cobb hold that an ethic aimed at maximizing richness of experience reflects the essence of evolution, for evolution exhibits a distinct trend toward increased complexity. God, who is Life itself, produces "realisation of value, that is rich experience or aliveness" (197). Therefore, they argue, life itself is purposeful. Humans, as the most complex life-form, are the ultimate product of this evolutionary drive. It is important to note that this argument provides an evolutionary justification for conferring the greatest value upon humans as the "supreme achievement of Life" (ibid.).

Yet the evolutionary basis for this interpretation of value is scientifically suspect. Not only do Birch and Cobb understand evolution as a complex yet linear process characterized by "the cosmic aim for value" (ibid.), they also, like Moltmann, lapse into non-Darwinian (often Lamarckian) language to describe the processes of evolution. Like Moltmann, they place great emphasis upon the direct role of environmental factors in molding organisms by provoking responses from them. Again, it is not apparent what role, if any, natural selection plays in this process of transcendent evolution.

Birch and Cobb argue that evolution exhibits a sort of pioneer spirit that maintains creativity in the face of changing situations. Animals contain within themselves the "main agency of evolutionary development" (ibid.). Evolution involves "emergent possibilities for creative transformation in each new situation" (188). Nature "actualises creative novelty" whenever it can, bringing "order out of chaos" (189, 192). What is "novel" in all life-forms, they argue, is the "response" of organisms to their environments. But natural selection explains this process differently: animals' particular (random) traits *may or may not* confer an advantage in a given environment. If they do, they are preserved.[69]

Birch and Cobb also make certain assumptions about evolution as necessarily and inherently *directed toward* increasing intelligence and richness. The idea that evolution manifests a discernible drive toward greater consciousness is not new in the history of evolutionary theory. To give just one example from the history of biology, around the turn of the nineteenth century, American paleontologist Edward Drinker Cope proposed such a theory of evolution. He argued that organisms respond to changes in their environments by an exercise of choice. Consciousness itself, he maintained, was the principal force in evolution. Cope credited God with having built into evolution a life force that propelled organisms toward ever higher levels of consciousness and, like Birch and Cobb, he also argued that in humans consciousness had reached a new plateau of awareness unknown to other organisms.

Cope, however, was not, strictly speaking, a Darwinian. "Deeply religious," he argued that "Lamarckism left room for consciousness to be seen as the guiding force in evolution." Like Birch and Cobb he also suggested that animals "control their own evolution" owing to conscious responses to environmental change.[70] The development of mind itself, he believed, was promoted by evolution favoring ever increasing intelligence, with humans "as the ultimate expression of this trend."[71] Cope did not deny that natural selection played at least a secondary role in evolution, but held that the selection process, as described by Darwin, could not account for the *origin* of the fittest.[72] He suggested, in contrast to Darwin's theory, that evolutionary fitness originated with the activity of the organism itself.

Like Birch and Cobb's description of organismal action in evolution, this theory stressed the influence of external factors in evoking evolutionary responses and development. Increasing consciousness, Cope believed, was inevitable since he assumed that "increased intelligence always gives a better response to the environment" and will therefore be favored.[73] Birch and Cobb present a strikingly similar, non-Darwinian account of evolution. They too claim that evolution aims at conscious experience and, to this end, generates life-forms and selects among them in such a way that some organisms "will emerge with greater intelligence and capacity for feeling."[74]

Although they are more cautious than Cope about positing a single, divine plan for all life, their understanding of the evolution of consciousness and environmentally induced response is similar. Indeed, they even identify God as the life force in evolution, the "life-giving principle," the very "embodiment of internal relations" of organisms to their environments.[75]

This stress upon the influence of the environment for organisms (which, as we have seen, is a central feature of Birch and Cobb's ecological model) is consistent with a Lamarckian perspective, which assumes a *direct* relationship between an organism and its environment. Lamarck argued that evolution generates increasing complexity, driven by a "force that tends incessantly to complicate organization."[76] Organisms themselves sustain this complicating force by responding creatively to the "refining fire" of their environments.[77] Again, as in Moltmann's account, this means that natural selection is rendered largely superfluous. God is associated with the efforts of living organisms to meet the challenges of their environments head on and, in so doing, evolve.[78] Birch and Cobb's repeated emphasis upon the ecological model's view of organisms as "constituted by the environment" does not necessarily have anything to do with Darwinian evolution.[79]

Some ecotheologians' insistence on organisms as subjects may in part contribute to their neglect of a more Darwinian view. The belief that organisms are responding subjects, refashioning themselves in accordance with changes in their environments (let us call this a "transformationist" account), fits better with a Lamarckian account than with a Darwinian one. As biologists Richard Lewontin and Richard Levins note, Lamarck's theory depicts the environment as "the impetus for the organism to change itself through willing and striving."[80] Classical Darwinian theory, which emphasizes random variation, is in many ways more consistent with a view of organism as *object*. "Darwin's variational theory is a theory of the organism as object, not the subject, of evolutionary forces. . . . The organism is the object of these internal forces, which operate independently of its functional needs or of its relations to the outer world." No wonder ecotheologians find the Darwinian view unpalatable. "In a curious way," Levins and Lewontin continue, "the organism . . . becomes irrelevant for the evolutionist."[81] While Darwinism casts organisms as objects, they concede that some biologists have taken this view too far, arguing that organisms are mere robots controlled by their genes (such a view surfaces among gene-centered theorists like Dawkins, who characterizes organisms as survival machines at the mercy of selfish genes). Levins and Lewontin go on to say that this *extreme* "object" view of organisms is the product of a vulgar or superficial Darwinism. They propose instead a dialectical view of the organism in which the "internal and the external factors, genes and environment, act upon each other through the medium of the organism." Though the organism remains a medium at the intersection of forces, it is not merely passive matter but rather "is also the subject of its own evolution."[82]

Ecotheologians might well agree with certain aspects of this dialectical account. But Levins and Lewontin's arguments raise an important point about the subject-object distinction in ecotheology. Ecotheology's neglect of Darwinism may be in part a reaction to the vulgar or superficial Darwinism Levins and Lewontin critique.[83] In other words, the *perception* that Darwin's theory promotes a view of organisms as mere objects may perpetuate a dislike for Darwinism as "mechanistic" and "Cartesian," resulting in failure to incorporate his theory, even if this perception is false. If some biologists (such as Dawkins) have overstated the robotic nature of organisms, ecotheologians such as Birch and Cobb may go too far in the other direction in their accounts of subjecthood. It is important to note that Levins and Lewontin do not reject the idea of organism as object altogether but merely temper this view by considering the ways in which organisms also interact with and alter their environments. Thus, they argue, Darwin's view of the organism was an "important step in shaking free of the Lamarckian transformationist model of evolution," but the process begun by the theory of natural selection will not be complete "unless the organism is reintegrated with the inner and outer forces, of which it is both subject and object." Simply equating evolution with the ability of organisms to generate new, intelligent responses to changing conditions, however, Birch and Cobb suggest that evolutionary adaptation is accomplished in a single movement. Selection, at most, merely ensures that their active responses are further reinforced and passed on, but this is an interpretation owing more to Lamarck than to Darwin.

Moltmann, Cobb, and Birch set out to explain the role of God in evolution, in terms that are consistent with current scientific data. From a scientific perspective, process thought initially fares better, owing to its understanding of suffering in nature as an inherent and necessary element of interdependence. But ultimately the virtues of this account are overshadowed by an ethic that places inappropriate emphasis on removing suffering and evaluates animals in light of a criterion—richness of experience—that remains anthropocentric rather than biocentric or ecocentric. Their account of the evolution of consciousness, like Moltmann's theory of ecological space, draws on evolutionary explanations that are not necessarily consistent with evolution by natural selection. Moltmann takes an overly adaptationist approach, directly crediting God with fitness in nature, while Birch and Cobb see individual organisms themselves as loci of positive transformation that is lured forward, more indirectly, by God.

As in the work of other ecotheologians, the desire to promote an ecological model of nature as a living community, in lieu of atomistic or mechanistic views, creates inconsistencies in their arguments about nature and ethics. In particular, these authors are uncertain how to treat suffering and discord in nature, given their own interpretation of ecology as the study of interdependent and interrelated communities. Whereas ecofeminists tend to discount the role of suffering in their descriptions of nature but endorse an ethic that aims at its removal, Birch and Cobb acknowledge the necessity of suffering directly but then back away from this realistic appraisal, invoking a liberation paradigm. Moltmann's approach lies somewhere in between process thought and ecofeminism in terms of his interpretation of suffering and conflict. In general, he overlooks these elements, emphasizing harmony and perfect adaptation. Yet suffering and conflict reemerge from time to time, primarily as a foil for the spirit of God which continually works to overcome the suffering and conflict that marred original creation. All these authors display a deep ambivalence about nature and the processes of evolution. The ecological model symbolizes an ideal natural community that ecotheologians believe, or perhaps merely hope, will one day become a reality, even if nature currently falls short of their expectations.

The focus on eradicating suffering, as well as a selective use of Darwinian data, is shared by secular environmentalists as well. The next chapter will focus on the work of two prominent philosophers, Tom Regan and Peter Singer, and their understanding of the moral status of nonhuman animals. In the work of these authors we will again encounter themes of liberation for other life-forms based on largely anthropocentric criteria that establish their common traits (sentience, intellectual and psychological faculties) with humans.

CHAPTER 4

Darwinian Equality for All

Secular Views of Animal Rights and Liberation

If a being suffers, there can be no moral justification for refusing to take that suffering into consideration. . . . The limit of sentience is the only defensible boundary of concern for the interests of others.

—Peter Singer, *Animal Liberation*

There are times, and these are not infrequent, when tears come to my eyes when I see, or read, or hear of the wretched plight of animals in the hands of humans. Their pain, their suffering, their loneliness, their innocence, their death. Anger. Rage. Pity. Sorrow. Disgust. The whole creation groans under the weight of the evil we humans visit upon these mute, powerless creatures. The fate of animals is in our hands. God grant us we are equal to the task.

—Tom Regan, "*The Radical Egalitarian Case for Animal Rights*"

In the present chapter we turn from a discussion of ecological theology and examine some important arguments in *secular* ethics regarding human obligations toward other forms of life. Animal rights and liberation arguments flourished in the 1970s in the wake of civil rights movements and a surge of interest in ecological issues. A link to civil rights concepts is apparent in Peter Singer and Tom Regan's arguments, both of which extend to animals certain ethical categories and assumptions traditionally restricted to human beings. The concept of rights, of course, has a long and rich history in Western social and political thought. Regan's argument is both modest and radical in its claim that this concept ought to include most—and perhaps all—nonhuman animals as well. Singer's ethics, and especially his concept of persons, also involves expanding traditional categories of human interests and capacities to include (most) animals. Thus Regan and Singer take an approach sometimes referred to as moral extensionism, and for both of these philosophers the extension of ethics to animals is predicated upon arguments about evolutionary continuity (what I will call interrelatedness or "genealogical interdependence") between humans and animals. The whole thrust of the critique of "speciesism"—which holds that different treatment of nonhuman animals based on their species membership is an unethical prejudice akin to racism—is derived from an evolutionary argument that humans and animals differ only in degree. Regan and Singer employ biological arguments to challenge what they see as our society's "speciesist" status quo.

I have chosen to focus on Singer and Regan partly because their views exhibit certain emphases—and weaknesses—similar to those of ecological theologians. Singer's argument entails a utilitarian analysis of the interests of nonhuman animals, while Regan adopts a deontological approach. But, aside from this important difference, there are many similarities between Singer and Regan, as there are between their arguments and those of certain ecotheologians. The parallels between Regan and Singer's arguments and those of Birch and Cobb are especially striking. The richness of experience criterion (and the account of evolution of consciousness that supports it) resembles Regan's arguments regarding animals as subjects of a life with inherent value. Value in animal (and human) life is similarly assessed according to the presence of certain mental and psychological capacities, although Regan is more radical in his extension of subjecthood than Birch and Cobb. Singer's arguments also draw on mental and physical properties—both intelligence and ability to feel pain—in evaluating the moral claims of individual animals.

Some important differences exist, of course, between ecotheologians' arguments and those of Singer and Regan. Most important, Singer is explicit in his rejection of a religious foundation for his ethics (Regan, as we will see, is less so). His analysis of animal liberation is completely devoid of ecotheology's interpretation of liberation as a religious (Christian) paradigm that is consistent with a love ethic and corresponds to a preference for the poor and oppressed. Also, ecotheologians tend to speak in broad terms of liberating and healing "life" in general or "nature" as a whole, whereas Singer and Regan typically focus on animals only, and often their concern is directed toward the plight of animals in very particular circumstances. Ecotheologians express much greater interest in, and concern for, the well-being of a large, ecological "community" of organisms or a "web of life" (although they may fail to understand why this focus is inconsistent with an ethic of liberation or care for each individual "subject" within that community.) For them interrelationship defined in terms of ecological interactions is as important as interrelationship defined as biological kinship.[1]

However, Singer and (especially) Regan are adamantly *un*concerned with the moral status of *larger* aggregates of beings such as species or ecosystems, whereas ecotheology makes frequent reference to community life and the good of its members. The ecological model and community interdependence are not prominent themes in the work of these philosophers. Nevertheless, a variety of the concept of interdependence runs through all these arguments, namely, interdependence as *evolutionary continuity* and the kinship of all life. Each of these authors contends that his understanding of life is informed by an evolutionary perspective; much of our discussion here will focus on the incomplete use of "evolutionary" information that is common in environmental arguments, religious and secular alike.

Singer and Regan employ a biological/evolutionary argument for condemning speciesism, but I contend that this refutation of speciesism is misapplied when combined with an ethic of reduced suffering or individual rights for animals. Put differently, these arguments tend to draw selectively from evolutionary theory, highlighting the role of evolution in producing similar *faculties* and *capacities* among humans and animals, while ignoring the implications of evolution that do not sit well with their arguments, such as the necessity of suffering in nature and the biological significance of the *species* level. Environmentalists are often happy to invoke a Darwinian account of our kinship and psychophysical similarity with animals in order to correct biases against animals; however, they prefer not to let an evolutionary perspective interfere with their belief in the rights of the individual organism and their characterization of suffering as an evil to be eradicated. Selective use of evolutionary theory presents problems that we will return to throughout the discussion of Regan and Singer that follows.

Having said that, I nevertheless think that there is value in the work of these philosophers, particularly Singer's. The extreme reaction that Singer's arguments have provoked may well be an indication of the huge gap that still exists between the inordinate concern in our society with any perceived threat, major or minor, to the sanctity of human life and the almost complete lack of concern for animal life. When Singer disparages the "striking contrast" between many people's outrage over the killing of severely handicapped and suffering newborns and the "casual way in which we take the lives of stray dogs, experimental monkeys, and beef cattle," I share his frustration.[2] The primary focus of my critique of Singer's ethics will be the implications of his work for the suffering and death of animals in nature and the anthropocentric assumptions embedded in his evaluation of animals and the environment as a whole. The aspects of his work usually considered the most controversial are not, for the most part, points that I find especially problematic.

The rift that exists between animal rights/liberation arguments on the one hand and holistic, ecocentric environmentalism (such as land ethics) on the other is also an important theme in the discussions that follow. As I have already suggested, Regan and Singer resist an ethical focus on aggregates and communities of beings. They acknowledge that animal rights and liberation arguments do not mix well with holistic approaches that seek to preserve a biotic community, sometimes at the expense of individual lives. Nonetheless, Singer and Regan attempt in different ways to apply their arguments to nature as a whole (that is, not just to animals in factory farms, laboratories, or other cultural contexts). Try as they might to formulate an environmental ethic, their arguments, like those of ecotheologians, remain largely incompatible with evolutionary realities because they focus on the humanlike capacities of animals (sentience, psychological complexity, subjecthood) and give primacy to individual organisms. We will see that even the concept of speciesism—which many animal advocates rely upon so heavily—is problematic in the context of environmental ethics, despite its apparent evolutionary rationale.

An additional reason for examining the arguments of secular ethicists, aside from certain similarities with ecotheology, has to do with the status of debates about animal rights and liberation within the field of environmental ethics as a whole. Ecotheologians' interest in the liberation (and sometimes "rights") of animals echoes arguments that have been issued by secular ethicists since at least the early 1970s. Rights and liberation arguments have consistently been met with harsh criticism by more holistic environmentalists. In fact, in the more than twenty-five years since Singer's *Animal Liberation* appeared, a significant portion of the debate in environmental ethics has focused on this controversy. For example, one of the leading journals in this area, *Environmental Ethics*, has featured essays about the incompatibility of animal liberation and environmental ethics since its inception; much discussion has been devoted to the difficulties of constructing environmental approaches upon a concern with animal sentience and rights.

The recognition that these movements are often at cross-purposes is as old as the field of environmental ethics itself. Yet this rather prominent debate within secular environmental ethics—between animal advocates and ecocentric environmentalism—has been virtually ignored among ecotheologians who, on the whole, continue to concern themselves with issues of animal suffering, sentience, and liberation.[3] Many ecotheologians, it seems, are not conversant in these important debates. Their continued preoccupation with extending ethical consideration to animals based on their possession of certain humanlike characteristics reflects a lack of familiarity with (or perhaps interest in) important developments within the broader discipline of environmental ethics to which they hope to make a contribution. If ecotheologians paid closer attention to these debates, perhaps they would be more cautious about reiterating similar, questionable arguments regarding the suffering and rights of individual animals in nature.[4]

We will begin with Singer's arguments on the issue of inflicting pain and the ethics of killing animals and humans. To anticipate one objection: Singer might very well reply that his arguments about animal liberation are not meant to include *all* animals; he is primarily concerned with factory farm and laboratory animals. Nevertheless, his arguments and classifications undeniably blur the distinction between domestic and wild, and he interprets the value of all kinds of animals from the standpoint of sentience and mental level. In his writings that deal directly with environmental issues, Singer does not attempt to modify his basic line of argument to fit animals in the wild. He simply supplements these arguments with other utilitarian considerations that might give value to nonsentient forms of life (e.g., trees, plants), suggesting that these may have instrumental value for humans in terms of recreation, medicinal purposes, or other uses. The focus on humanlike properties among animal species, combined with a utilitarian assessment of nature's use value to humans, raises the question of whether such an ethic succeeds in escaping the very speciesist biases it seeks to eradicate. We will return to this issue shortly. For now, we will begin by examining Singer's account of liberation and the concept of speciesism.

ANIMAL LIBERATION

Singer is one of the most controversial thinkers of our time. In recent years his appointment as a bioethicist at Princeton University has sparked protests from critics (particularly pro-life activists) who accuse him of devaluing human life and promoting infanticide. Because of his claim that even lower species of animals should receive at least as much ethical consideration as human infants (if not more), Singer has garnered praise from animal rights activists and received severe condemnation—even death threats—from other factions.

Soon after Singer's book *Animal Liberation* appeared in 1975, it became the handbook of animal activists seeking to expose the evils of speciesism in all its forms. Despite his popularity among animal rights proponents, however, Singer's argument is not strictly speaking a *rights* position. The case for liberating animals can be made "without getting embroiled" in talk of rights at all, he argues.[5] In fact, as we will see, animal rights philosopher Tom Regan criticizes Singer's work on the grounds that it does not support equal rights for all animals. Still, Singer's concept of a "person" and Regan's definition of a "subject-of-a-life" are similar in many respects, and both arguments have engendered criticism from other ethicists who see such categories as irrelevant for environmental ethics.

SPECIESISM

Speciesism was coined by British animal activist Richard Ryder in the 1970s, but Singer's work has done the most to popularize the term, at least in the United States.[6] Speciesism refers to a bias against nonhuman animals simply because they are members of another species. Critics of speciesism such as Singer argue that species membership is not a *morally relevant* basis for excluding other living beings from the sort of ethical consideration we typically extend to humans. In this sense it is akin to racism, its critics hold, in that it permits systematic oppression of animals on grounds that are philosophically—and even biologically—untenable.

Speciesists (which, as far as its critics are concerned, is a category that encompasses most people) often defend their position toward nonhuman animals by pointing to some alleged characteristic of human life that makes it unique or "sacred." Human uniqueness may be understood in secular or religious terms; the most common defense of the uniqueness of humans on religious grounds is the belief that humans alone are created in the image of God and act as God's representatives on earth.[7] Secular defenses of speciesism have deep roots as well, and these can often be traced to Cartesian arguments that animals do not possess minds (as demonstrated by their lack of reason and/or language) or do not feel pain as humans do. Critics of speciesism contend that all arguments for radical separation of humans and animals, whether religious or secular, are undermined by the fact of evolutionary continuity among all living things, as well as the mounting evidence of the intelligence and complex emotional lives of animals. Darwin's work in evolutionary theory and his study of the similarities between human and animal emotions has—or *ought* to have—undercut the speciesist's case. Speciesists are left with no relevant criterion for excluding animals from moral consideration, critics argue. The biological continuum (genealogical interdependence) of humans and other animals implies a *moral* continuum, Richard Ryder argues, but speciesism "denies the logic of Evolution."[8] As we have seen from the arguments of ecotheologians, the profound interrelationship of all life-forms, and the ethic of love or liberation that interrelationship implies, represents a similar attempt to undermine Cartesian dualism on the basis of biological continuity.

Following Ryder, Singer argues that we should condemn oppression of animals for the same reasons that we condemn oppression of certain classes of people, most notably racial minorities (a point that seems to be lost on critics who accuse him of advocating genocide).[9] A liberation movement in Singer's view involves a "demand for an end to prejudice and discrimination based on an arbitrary characteristic like race or sex."[10] By the same token, *animal* liberation entails the removal of a prejudicial attitude that has persisted—when all other arguments have failed—simply because they are members of different species. Singer's understanding of animal liberation largely concurs with that of Birch, Cobb, and ecofeminists who also liken mistreatment of animals to oppression of humans on the basis of race and gender. Birch and Cobb, for example, argue that "those without power, be they women or Blacks or other oppressed groups are too easily denied agency as subjects. . . . It is from the results of just such objectification, of regarding people and other living organisms as object only, that many are now seeking liberation."[11] Recall that McFague, too, understands nature to be "the new poor," an oppressed class that "commands our special attention," just as other oppressed classes have.[12] On the whole, *liberation* signals for all these authors the removal of conditions of suffering, recognition of subjecthood in all beings, and an end to the objectifying oppression inflicted upon both human and nonhuman forms of life. Let us look more closely at Singer's utilitarian account of inflicting pain, killing, and his concept of personhood.

INFLICTING PAIN AND TAKING LIFE

Singer's ethics distinguishes (as much as possible) the issue of inflicting pain from the issue of killing.[13] Animal and human faculties relevant to the issue of taking a life are not necessarily relevant to the issue of inflicting pain. The existence of certain capacities—self-awareness and meaningful relationships, the ability to anticipate and plan for the future—have direct bearing on the issue of taking life. But pain is pain, Singer argues, regardless of the species of being that is experiencing it.[14]

Given the relevance of mental level for taking a life, it is often, but not always, the case that we would choose to save a human life over a nonhuman life, if such a choice had to be made. Singer denies, however, that giving preference to the human in such instances is a speciesist move because we must also recognize that the lives of certain animals are worth more than the lives of certain humans: "A chimpanzee, dog, or pig, for instance, will have a higher degree of self-awareness and a greater capacity for meaningful relations with others than a severely retarded infant or someone in a state of advanced senility."[15] When Singer argues that all animals are equal, he really means that all *sentient* beings are entitled to equal *consideration* of their interests. But different kinds of species have different sorts of capacities. This, in turn, implies that they have different interests. Once we weigh these interests, we may judge that killing a certain creature is more wrong than killing another. This is not an a priori speciesist judgment because such decisions are made with reference to the particular capacities involved in each case, not with reference to species membership.

With the issue of pain, the matter is simpler. Although evidence for animal rationality exists (particularly among primates), these factors do not determine the wrongness of inflicting pain. Singer sees the capacity to suffer as a sufficient condition for including any being within one's circle of moral consideration. "The capacity for suffering and enjoyment," Singer writes, "is a *prerequisite for having interests at all*, a condition that must be satisfied before we can speak of interests in a meaningful way."[16] His approach to ethics is rooted in the thinking of classical utilitarian philosophers. Singer's arguments closely resemble those of John Stuart Mill and Jeremy Bentham insofar as his work supports the extension of moral sympathies beyond our own species; like them, he also places critical importance upon the issue of pain. Any being who is capable of feeling pain has an interest in avoiding it, regardless of mental sophistication. The crucial question regarding our treatment of animals, Bentham argued, is not "can they *reason*, nor Can they *talk*, but Can they *suffer*?"[17] Moreover, Singer, like Bentham, has turned his philosophy toward active reform—not mere analysis—of our treatment of animals.

Regarding the most common (cultural) sources of animal suffering—farming/meat production and animal experimentation—weighing the suffering experienced by animals against the enjoyment or benefits for humans indicates that these practices cannot be justified, Singer argues. The "simple, straightforward principle of equal consideration of pain or pleasure is a sufficient basis for identifying and protesting against all the major abuses of animals that human beings practice."[18] For example, the majority of animal experimentation inflicts far more suffering than it prevents, Singer claims, because most tests on animals serve no vital medical purpose. If it *were* the case (and Singer argues that it is not) that animal experimentation produced dramatic life-saving results for humans, then it might be justified, even though it involves suffering of animals: "If one, or even a dozen animals had to suffer experiments in order to save thousands, I would think it right and in accordance with equal consideration of interests that they should do so."[19] But since the vast majority of experiments do not produce such results, experimentation is highly problematic. Furthermore—and this is a crucial point in Singer's argument—even animal experiments that produce a great good for others cannot be justified unless we would also allow similar experiments to be performed on certain classes of humans. Singer asks, "Would experimenters be prepared to perform their experiments on orphaned humans with severe and irreversible brain damage if that were the only way to save thousands?"[20] If not, their use of animals constitutes speciesism. In such cases there is no morally relevant trait possessed by humans that is lacking in animals.

Meat eating, too (particularly eating the flesh of animals raised in the factory farm system), involves considerable suffering for animals, whereas human interest in eating meat (primarily because humans enjoy the taste of it) is a comparatively trivial interest that does not justify the suffering it entails. If all humans were like "Eskimos living in an environment where they must kill animals for food or starve," a diet of meat might be justified because our interest in surviving would not be a trivial one.[21] For most of us, however, our enjoyment of the taste of animals (or the deprivation we would experience in giving up meat) is outweighed by the interests of animals to avoid great suffering.

Regarding animals used in experiments, Singer is aware that his arguments can be interpreted in two ways: he may simply be urging that we dispense with experimenting on animals altogether, since most people would not condone treating humans in this way. Or, he could be interpreted as arguing that we may continue to experiment on animals *if* we allow similar treatment of certain classes of humans. Singer is not always clear on which of these views he means to endorse, and the second interpretation is, clearly, quite controversial. The crucial point is that "we should give the same respect to the lives of animals as we give to the lives of those humans at a similar mental level."[22]

Singer acknowledges that humans may anticipate pain in ways that animals generally cannot and thus are sometimes assumed to suffer more. However, he counters that we cannot use this fact as a reason for differential treatment of animals and humans *unless* we would also condone the same treatment of certain humans such as infants or the mentally retarded who may also lack an anticipatory dread of pain.[23] Furthermore, he suggests that animals may actually suffer more than (normal) humans *because* of their limited comprehension of what is happening to them. When I take my cat to the vet, I cannot explain to him (as I could to an adult human and children of a certain age) that the painful procedure he is about to receive is for his own good. Nor can my cat prepare himself mentally for the coming ordeal. In other words, lack of mental sophistication may intensify suffering. In any case, ability to suffer is more important than mental level, as far as inflicting pain is concerned. However, because animal experimenters often point to mental differences between animals and humans in justifying their work, they put themselves in a position in which they must either accept similar treatment of certain humans or be guilty of a speciesist bias.

Singer does not deny that experiences of pain vary from one organism to another: slapping a baby and slapping a horse (the example he uses) are not equivalent because the baby has thinner skin, is more sensitive to pain, and suffers more. Therefore we have to compare "like" kinds of pain. Aside from this qualification, Singer maintains that "pain is pain," regardless of who is experiencing it, and that pain is in itself bad. "How bad a pain is depends on how intense it is and how long it lasts, but pains of the same intensity and duration are equally bad, whether felt by humans or animals."[24] The principle of equality demands that all (like) suffering be counted equally "no matter what the nature of the being."[25]

KILLING: PERSONS AND NONPERSONS

Singer adds a new dimension to classical utilitarianism insofar as he focuses particular attention on what it means to be a *person*. This feature of Singer's utilitarianism has also engendered the bulk of criticism.[26] According to this view, some members of our own species are not persons and some members of other species are. Singer's understanding of what it means to be a person resembles other ethicists' accounts of humanness in that it stresses, for example, capacity to relate to others and a sense of one's past and future. However, Singer isolates two crucial characteristics—"rationality and self-consciousness"—as capacities worthy of special consideration for personhood.[27] He contends that being a person may be distinct from being a member of the species Homo sapiens. There are nonhuman (animal) persons as well as nonperson humans. A *person* is someone who is not merely conscious but self-conscious. Persons have an awareness of being distinct selves, with a sense of past and future, and at least some degree of rationality. Singer concedes that it is "notoriously difficult" to say with certainty whether or not another being is self-conscious, and therefore he does not draw a sharp line but argues that some beings are clearly persons and some probably are. In any case, he argues, when there is uncertainty regarding personhood, we should "give that being the benefit of the doubt."[28]

A stronger prohibition holds for killing persons than killing nonpersons, whatever the species. This means that one can make a better case against killing some animals than against killing certain humans: "Killing, say, a chimpanzee is worse than the killing of a human being who, because of congenital intellectual disability, is not and never can be a person."[29] By the same token, a human fetus is not a person and has no "claim to life" as a person. Indeed, even a newborn is not truly a person, according to Singer's definition, because it is neither rational nor self-conscious. Killing an infant, therefore, is not as bad as killing some animals. The life of a newborn is "of less value to it than the life of a pig, a dog, or a chimpanzee."[30] Singer's admission that his views of personhood and killing render some cases of infanticide permissible constitutes the most controversial aspect of his ethic.

Many animals are not self-conscious persons but are nevertheless "conscious" beings (fish and perhaps chickens, Singer suggests, fall into this category). In the case of animals that are conscious but not obviously persons, the ethics of killing them rests on "utilitarian considerations" rather than a basic "right to life" that persons have (169). Killing of nonperson animals must take into account both the pain inflicted in the process and the impact (grief, distress) on other animals around it. Such factors "would lead the utilitarian to oppose a lot of killing of animals, whether or not the animals are persons." It is not clear, however, how Singer can maintain that nonself-conscious nonpersons can nevertheless be capable of an experience such as "grief."

Among animals who appear to have qualities of personhood, the case against killing is made "most categorically" regarding "chimpanzees, gorillas, and orangutans" (132). A case can also be made (though somewhat less forcefully) against taking the lives of a second category of animals that includes "whales, dolphins, dogs, cats, pigs, seals, bears, cattle, sheep," and perhaps all mammals, Singer argues (ibid.). Finally, a weaker case can be made against killing animals that are not persons, i.e., nonself-conscious, nonrational animals. The wrongness, if it is wrong, of killing these animals stems from the loss of pleasure that killing them would involve. "Where the life taken would not, on the balance, have been pleasant, no direct wrong is done," Singer argues. Even if its life would have been pleasant, it is possible that "no wrong is done if the animal killed will, as a result of the killing, be replaced by another animal living an equally pleasant life" (133). Singer does not elaborate on the types of animals that comprise this category of nonpersons, but we can assume that it includes some species of fish and perhaps other small, nonmammalian creatures such as reptiles or amphibians. Singer acknowledges that this analysis makes it permissible to kill some animals under certain conditions (for food, for instance), but he concludes with a pragmatic argument for not doing so for the sake of "foster[ing] the right attitudes of consideration for animals, including non-self-conscious ones" (134).

These features of Singer's ethics—the issue of pain, killing, and personhood—suggest certain difficulties, particularly when his argument regarding the pain and moral status of animals is applied to animals in situations other than factory farms and laboratories. Let us consider these in turn.

ANIMAL LIBERATION AND ENVIRONMENTAL ETHICS: PROBLEMS WITH SINGER'S ACCOUNT

There are a number of potential problems with Singer's case for animal liberation. I want to focus primarily on his understanding of pain as an evil to be reduced or eradicated, his concept of persons and nonpersons, and (later in this chapter) questions raised by the concept of speciesism. As we will see, Singer's utilitarian emphasis on pain is at odds with evolutionary and ecological considerations, owing to a lack of consideration for the important, biological function of pain in Singer's account and the impracticality of defending individual animals in an ecological context. Similarly, we will consider why Singer's account of what it means to be a "person" cannot easily be reconciled with a workable environmental ethic because of two problems: 1. a failure to distinguish between the ethical treatment of wild and domestic animals and 2. the value his ethic places on animals that possess the most humanlike capacities. While it may not necessarily be an anthropocentric judgment simply to point out that animals who have similar physiology to humans probably have some similar capacities, it becomes an anthropocentric argument when the claim is made that presence or absence of these capacities determines the *value* of animals. As I hope to show, Singer draws on biological arguments about the human-animal continuum to make a case for their ethical treatment, but the resulting ethic cannot be considered nonanthropocentric. We will look first at Singer's account of pain and the problems it raises in an ecological, evolutionary context.

Pain Is Pain?

Making the capacity for suffering the "vital characteristic that entitles a being to equal consideration" creates problems for an ethic extended to animals as a whole (57). Singer suggests throughout his arguments that pain is pain regardless of the species of being who experiences it, yet at various points he seems to equivocate on this issue.

If a being is suffering, Singer argues, "there can be no moral justification for refusing to take that suffering into consideration" (ibid.). Does Singer believe, then, that we are obligated to consider (and even reduce) the suffering of all animals, including those in the wild? He does not write at length on the status of wild animals in *Animal Liberation*, which is geared primarily toward speciesist practices involved in factory farms and laboratory research on animals. However, he does state that a reconsideration of the suffering of other beings would force us to make "radical changes" in the way we treat animals in a variety of contexts, including "our approach to wildlife."[31] Some of Singer's other works contain more direct discussions of environmental issues and we will look at a few of these as well.

Singer is aware that at least some of the suffering that animals experience is "natural"—animals eat other animals, they are exposed to extremes of temperature, hunger, and disease. How is natural suffering to be reconciled with an ethic of reduced pain for animals? Singer does not dwell on this point but grants that an ethic based upon the reduction of the total amount of suffering seems to imply the elimination of carnivores. "It must be admitted," he argues, "that the existence of carnivorous animals does pose one problem for the ethics of Animal Liberation, and that is whether we should do anything about it."[32] Surprisingly, in response to the "carnivore problem," Singer appears to advocate a noninterventionist ethic toward wild animals,[33] replying that humans should refrain from playing the role of tyrant over animals, whether in the laboratory or in nature. "We should leave them alone as much as we possibly can."[34] On the other hand, he acknowledges that attempts to rescue whales trapped in Alaskan ice are morally commendable; only a "callous person" could argue otherwise. Eliminating natural carnivores is different from saving trapped whales and, in general, we should "not try to police" nature.[35]

Singer rejects the argument offered by some critics of animal liberation that the suffering animals experience in factory farms is often less than that experienced by wild animals. He counters that, even with all its attendant dangers, a life in the wild is a life of "freedom," and "surely the life of freedom is to be preferred," despite the presence of suffering. Factory farm animals endure a life of "utter boredom," by contrast.[36] It should be noted, however, that this comparison between the pain of wild and farm animals points to the fact (a fact he does not pursue further) that there are different *kinds* of pain. When comparing the pain of animals in the wild to those in factory farms or laboratories, Singer's remarks indicate that the main issue is not (as he has previously maintained) how intensely or how long an animal experiences pain but what *role*, if any, pain plays in the overall existence of the animal. The physical pain that is suffered (or at least risked) by a wild animal is outweighed, Singer seems to suggest, by the overall experience of wildness and freedom. Freedom for wild animals does not mean freedom from pain: "Many wild animals die from adverse conditions or are killed by predators."[37] Thus life in the wild is good, he seems to think, even though it involves pain. But if this is the case, then it is not true (as Singer asserts throughout *Animal Liberation*) that "pain is pain" and that "pain and suffering are in themselves bad" regardless of "the nature of the being" in question. In fact, one might even assert that pain in such an instance is in some sense *good*.

Singer equivocates on the "pain is pain" argument in other places as well. Despite his statement that we should support an ethic of reduced suffering for all animals, he sometimes alludes to the "evolutionary value of pain."[38] Like other environmentalists, Singer is interested in the potential of an evolutionary perspective for heightening our awareness of the pain of other animals who are our kin. In making the case that other animals experience pain in much the same way humans do, he embarks on a discussion of the evolution of the nervous system: "It is surely unreasonable to suppose," he writes, "that nervous systems that are virtually identical physiologically, have a common origin and a common evolutionary function, and result in similar forms of behavior in similar circumstances should actually operate in an entirely different manner on the level of subjective feelings."[39] In the course of making this argument, he also observes that "capacity to feel pain obviously enhances a species' prospects of survival."[40] If pain has evolutionary value for animals in the wild then, again, it seems that pain is *not* pain—that is, we should not treat all pain (even "like pain") in the same way, and certainly we should not always consider it "bad in itself." An animal in the wild has a similar nervous system to one in cultural contexts (say a wild boar as compared to a pig in a factory farm). But surely we do not have the same obligation to reduce the boar's pain, regardless of its intensity or duration. Singer does not really address these differences, nor does he address the implication that our obligations to animals differ in wild nature. His general policy is that we should leave nature alone, but, as we will see when we turn to Regan's ethics, this is an insufficient answer.

The existence of evolutionary value of pain implies that we should *not* assume that all animals who can feel pain have an interest in avoiding it, as Singer has previously stated. Indeed, animals may even have an interest in *not* having their pain reduced. But perhaps Singer would reply that I have misinterpreted his point here: maybe he is simply saying that even animals in the wild have an interest in avoiding pain—otherwise they would not run from predators or avoid other situations that were likely to bring about pain or death. As an empirical fact, their behavior shows a preference for not being in pain. This is certainly true. But it is one thing to say that animals will avoid pain on their own; it is another to suggest that our responsibilities toward them are generated by their interests in avoiding pain. The fact that animals avoid pain in the wild does not create an obligation on our part to help them do so.

J. Baird Callicott has criticized animal liberation arguments along these very lines. "If nature as a whole is good," he argues, "then pain and death are also good."[41] What makes laboratory testing and factory farming of animals immoral is not the pain inflicted upon them but the "transmogrification of organic to mechanical processes."[42] That is, the issue is not whether to eat vegetables or animals but whether to eat organic foods over "mechanico-chemically produced" plants and animals. The preoccupation with animal pain is beside the point.

Callicott has since changed his opinion of vegetarianism, arguing that holistic environmental ethics does in fact call for a nonmeat diet (largely because of the impact on the environment of raising food animals). However, his main arguments, including his comments about the role of pain in the life of *wild* animals, still stand. One of the "most disturbing implications" drawn from animal liberation theory such as Singer's, he argues, is that, "were it possible for us to do so, we ought to protect innocent vegetarian animals from their carnivorous predators."[43] Singer's statements that "if a being suffers, there can be no moral justification for refusing to take that suffering into consideration" seems to encompass *all* living creatures, wild and domestic, despite the noninterventionist policy toward wild animals hinted at elsewhere.

Perhaps Singer is arguing only that we should reduce the pain that we *inflict* on animals, not all pain wherever and however it occurs. He might respond that the only cases in which we should concern ourselves with animals in nature would be those instances where our activity (such as clear-cutting a forest or building a dam) leads directly to suffering or death of animals who would otherwise experience less suffering or more pleasure. In *Practical Ethics* Singer deals with such examples. Let us look at one case of human-inflicted suffering in order to see how Singer's ethic of minimizing suffering of sentient beings applies to nature.

The central problem with an argument such as Singer's—a problem he readily acknowledges—is that it seems to offer no grounds for protecting nonsentient nature, whether individual trees and plants or tracts of wilderness that do not have interests and cannot experience pain. In order to account for these, Singer points to the value of nature as a resource for meeting the needs of sentient beings. According to a utilitarian perspective, wilderness "is valued as something of immense beauty, as a reservoir of scientific value still to be gained, for the recreational opportunities it provides," and perhaps because people "just like to know" it exists.[44] Thus, he argues, a "human-centered ethic can be the basis of powerful arguments for what we may call 'environmental values'" so long as those values are not strictly economic, since economic considerations usually do not take into account the interests of nonhuman animals.[45] Before building a dam, for instance, we should consider not only the benefits to humans but also the suffering and death caused to sentient animals whose habitat it would destroy.

But what about the nonsentient beings themselves? Are there grounds for valuing plants or insects that are threatened by human activities but have no known utilitarian value to anyone? Singer acknowledges that the problem is a difficult one because only certain kinds of entities—sentient individuals—can be said to have interests. Holistic environmentalists such as land ethicists and deep ecologists have attempted to articulate the value of nonsentient nature, he notes, but these approaches do not impress him. Holistic or ecocentric approaches make no sense ethically, he argues, because they tend to value a larger whole such as a species or ecosystem.[46] We cannot talk about "interests" of a species apart from the combined interests of each sentient individual. "Species as such are not conscious entities and so do not have interests above and beyond the interests of the individual animals that are members of the species."[47] The combined interests of a plant species is still zero, since individual members are not sentient. Nonsentient individuals such as trees and larger aggregates such as ecosystems have the same moral status—or lack of moral status—according to Singer's approach. "Trees, ecosystems, and species are more like rocks than they are like sentient beings" in not possessing interests and therefore not generating obligations on our part. All we are left with, then, as an ethical basis for preserving wilderness is "arguments based on the interests of sentient creatures, present and future, human and non-human," and these, Singer maintains, are "sufficient" for preserving our remaining wild places.[48]

Putting aside for the moment the question whether instrumental value is sufficient to preserve nonsentient aspects of nature, it is not clear that Singer's approach can even protect *sentient* beings. One of the problems with an ethic based upon counting the interests and minimizing the suffering of sentient beings is that it may turn out that the best way to reduce the total amount of suffering would be to take sentient animals out of their environments altogether and use their natural habitats for our own enjoyment. Philosopher Bryan Norton points out that destruction of a species' habitat might lead to an overall reduction of suffering. We might "humanely capture and relocate" all animals in an area to be cultivated for human use, thereby removing sources of suffering, particularly if animals were transferred to zoos where they would be fed, sheltered, and protected from predators.[49] In any case, it is not implausible that minimizing suffering and permitting environmental destruction can be maintained simultaneously.

It seems more reasonable to approach environmental issues with a different ethic altogether rather than trying to stretch the idea of interests of sentient beings (combined with enlightened self-interest) to cover all kinds of situations involving nonhuman life. Singer is right to think that there is something odd about weighing the interests of ecological entities such as ecosystems or species. It is true that these species have no "moral agency, reflective self-awareness, sentience, or organic individuality."[50] But the oddness of such an approach stems from an environmental perspective based in these *capacities*, not from ecological concepts themselves. As Rolston argues, it makes perfect sense to talk of our duties to species so long as we understand that our obligations are not so much to the species itself as to the *process* of speciation:

> What humans ought to respect are dynamic life forms preserved in historical lines, vital informational processes that persist genetically over millions of years, over-leaping short-lived individuals. It is not *form* (species) as mere morphology, but the *formative* (speciating) process that humans ought to preserve.[51]

Respect is owed not to a label, class, or category but to forces in nature, and such forces do not necessarily respect the lives of individuals. Rolston cites the phenomenon known to biologists as "genetic load" to illustrate why preserving a natural *process* conflicts with an ethic that aims primarily at the preservation of individual lives. Owing to variation, detrimental genes that can reduce the health or fertility of a species are often distributed throughout the entire gene pool of the species. Some members of the species will turn out to have a greater load of these genes than others. "Less variation and better repetition in reproduction would, on average, benefit more individuals in any one next generation, since individuals would have less 'load.'"[52] But this is not how evolution works. From the perspective of the individual with detrimental genes, this process has obvious disadvantages. But what is bad for the individual can be advantageous to the species because variation provides more overall stability and better chance of survival of the species *as a whole* in changing environments.

Other examples illustrate the ecological significance of species. Forest fires are necessary for the continuation of some species of trees such as aspen. Park officials refrain from putting out natural fires (and sometimes start the fires themselves) because they are an important part of forest succession, which the threatened species needs to survive. The fires cause harm, even death, to individual organisms—including individual aspen themselves—but they preserve the lifeline of the species as a whole. Thus the "good" of the species and the "good" of the individual are not necessarily equivalent.

Humans should not attempt to alter this scenario, regardless of the suffering it causes to individual beings.

The same evolutionary processes that led to the sophisticated nervous system that Singer so respects in individual, sentient beings also preserves species, not the individual, over time. An environmental ethic should attach value to the processes of evolution, not merely their products. Natural forces themselves suggest the more appropriate ethic. Singer's ethic merely locates humanlike categories of experience in certain parts of nature and attempts to value the remaining parts in terms of human interests. "To value all other species only for human interest is like a nation's arguing all its foreign policy in terms of national interest," Rolston argues. "Neither seems fully moral."[53]

This discussion of individuals and species returns us to a second feature of Singer's ethic—the issue of killing (distinct from causing suffering to) individual beings. Clearly, some of the same objections to his argument would apply, whether the case involves suffering or death, since many conditions in nature that cause suffering also result in death. Recall that Singer understands the ethics of killing to be connected to mental level, self-consciousness, and what he calls "personhood." Does this concept apply to animals in natural environments as well? Let us return to this concept and see what it implies for animals in nature and our ethical obligations toward them.

PROBLEMS WITH SINGER'S CONCEPT OF PERSONS FOR ENVIRONMENTAL ETHICS

Singer does not consistently distinguish animals in wild and nonwild contexts. According to his criteria, the greatest value is always assigned to animals possessing personhood; as we have seen, it is "most categorically" wrong to kill persons (having rationality and self-consciousness). According to preference utilitarianism, killing a person who is "highly future-oriented in their preferences" is far worse than killing nonpersons who do not conceive of themselves as having a future, cannot anticipate and plan, and thus cannot be said to have preferences.[54] In this sense, it is much less problematic, and perhaps even permissible, to kill nonpersons, whether human or animal. "It would not be speciesist to hold that the life of a self-aware being, capable of abstract thought, of planning for the future, of complex acts of communication, and so on, is more valuable than the life of a being without these capacities."[55]

Note that each of Singer's categories of 1. persons, 2. possible or semipersons,[56] and 3. nonpersons include *both* wild and domestic animals. Primates (apes, in particular) rank highest among animal persons, in terms of intelligence and self-consciousness, Singer argues. Their lives have the same value and "claims to protection" as those of human persons.[57] Killing them, whether in experiments, for food, or in the wild, is extremely difficult to justify. Killing other animal persons or probable persons (a category that includes a mix of domestic animals such as dogs and cats as well as wild animals such as dolphins and whales) is difficult to justify, according to Singer.[58] Because Singer argues that, when in doubt, we should err on the side of assuming personhood, he would consider killing these "possible" persons very problematic. What does Singer's argument here imply for our treatment of animals in the wild? It is not unusual for environmentalists to have to make decisions that involve culling one species in order to preserve another that is threatened or endangered. What happens, for instance, when a conflict arises between animals that are merely possible persons (or probable *non*persons) and animals that are clearly persons? Or what about a conflict between an organism that is not even a conscious nonperson (such as a plant) and one that is a person or semiperson? The second question brings us back to the problem of what value, if any, nonsentient organisms such as plants have.

Let us look at one such case involving a population of rabbits and rare plant species, *Dudleya traskiae.*[59] The National Park Service killed or removed hundreds of rabbits and other exotic herbivores on Santa Barbara Island in order to preserve a small number of unexpectedly discovered plants that had been listed as extinct in the early 1970s. Why would a few plants have greater value than a large number of rabbits, which, if not self-conscious are at least conscious (and certainly can feel pain). The reason, again, lies not in the mental levels or other attributes of these organisms but in the value of species as opposed to individuals. Killing individual rabbits does not extirpate the species, whereas allowing the remaining plants to die would almost certainly mean the end of the *species* (whose common name, ironically, is *Liveforever*) as well as the individuals. Decisions in wildlife management do not hinge on the presence or absence of mental capacities or ability to feel pain. The Endangered Species Act, for instance, assumes that the "biological differences between animals and plants underlying their taxonomic separation offer no scientific reason for lesser protection of plants."[60]

There is very little reason, in Singer's ethic, to be concerned with preserving these plants—unless, of course, they turned out to have great value to us (say, medicinal value). But even if such use value can be ascertained, it is not clear that these values are sufficient to ensure protection. Quite the contrary, some plant species would be far better off, in terms of their overall chances of survival, if they had no known value.[61] Since herbal supplements and remedies have become popular, wild populations of certain plants that once grew abundantly, such as ginseng, are now endangered. In parts of Canada, more than 80 percent of wild ginseng populations have been decimated and are currently at risk of extinction.[62] Laws can be passed to declare their endangered status, but there will always be "poachers" who continue to harvest wild plants rather than putting the money and effort into cultivating crops. (This is also the situation with *Dudleya traskiae*, mentioned previously, which is prized by collectors.)[63] By the time such species are granted legal protection, their wild populations may have dropped to levels from which the species cannot recover.

In short, Singer assigns very low value to a large class of organisms: animals with little or no sentience and all plant species. When extended to all animals, this argument suffers from the same problems as Birch and Cobb's richness of experience criterion. Essentially Birch and Cobb's category of organisms possessing only instrumental value (those with little or no richness of experience) overlaps with Singer's category of nonpersons (though they draw lines somewhat differently between those they consider highly valuable). Whenever these conflicts arise, Singer, like Birch and Cobb, would give preference to the lives of animals who are most humanlike in their capacities—those capable of abstract thought, communication, complex social arrangements, etc. Birch and Cobb's reasons for taking porpoises and dolphins seriously consist in a "high degree of communication . . . deep emotions . . . [and] relatively abstract thinking."[64] Singer cites similar capacities as evidence of the wrongness of killing animal and human persons. While he denies that giving preference to beings with these qualities is a speciesist bias, since preference does not necessarily fall along the species boundary, this interpretation of value has not fully escaped the biases he condemns so long as his ethic provides little ground for valuing the lives of beings that are not like (normal) humans in any obvious ways.

Regan too opposes a speciesist approach to valuing animals and, like Singer, uses evolutionary evidence to show that preferences along species lines are biologically and morally untenable. Regan's ethic also makes moral concern for ecosystems or species as a whole difficult to defend—a point that he, like Singer, readily concedes. Nevertheless, he does not attempt to amend his argument to account for ecological considerations, nor does he refrain from applying his rights argument to animals in the wild. Although he and Singer are generally more interested in treatment of animals in nonwild contexts, they both maintain that their ethics can coherently apply to animal life as a whole.

TOM REGAN'S CASE FOR ANIMAL RIGHTS

Regan's arguments resemble Singer's, despite his emphatic rejection of a utilitarian analysis of animal ethics. Singer's concept of persons and Regan's definition of subjects of a life invoke many of the same kinds of characteristics that render animals worthy of moral consideration. Both philosophers focus on moral consideration of individual animals (as opposed to species) as well. As such, both perspectives are vulnerable to criticisms from holistic environmental ethicists who view their arguments as incompatible with ecologically oriented ethics.

Despite the frequent association of Singer's work with an animal rights agenda, Regan is a more appropriate representative of this approach. He builds his argument upon a human model of rights, claiming that the same grounds for granting rights-bearing status to humans can be applied to most and perhaps all animals. He presents a deontological argument that humans and animals have rights by virtue of being ends in themselves. Humans have value inherently, Regan argues, *not* just because, and not just so long as, they are good for something or somebody else. In most cases, the rights of the individual "trump the goals of the group"[65] and may be overridden only in unusual cases.[66] Mistreatment of beings with intrinsic value entails treating them as if they had only instrumental value. Furthermore, Regan argues, all individuals who have inherent value have it equally, and therefore all have the same basic rights.

Regan rejects Singer's utilitarianism, which denies *equal* inherent value. Equality for Singer means equal consideration of interests, but Regan has a more robust notion of equality in mind. The problem with Singer's ethics, he maintains, is that it does not show *why* it is wrong, either in principle or on utilitarian grounds, to subject animals to conditions of factory farms and laboratory tests. No necessary connection exists between the equality of interests principle and the utilitarian principle of maximizing the balance of good over bad. We might count animals' interests equally with those of human individuals and yet favor human interests on the grounds that we are bringing about the greatest possible balance of good over evil.[67] Current "speciesist" treatment of animals could be justified (as could racism and sexism) if it turns out that such treatment leads to better consequences, on the whole, than alternative treatment of animals. Singer fails to demonstrate that "*not* raising animals intensively or *not* using them routinely in research leads to better consequences, all considered, than those which now result from treating animals in these ways."[68] Speciesism could be justified on utilitarian grounds. Regan proposes instead an ethic of "radical egalitarianism" for animals as subjects with evolved capacities of consciousness.

EVOLUTION AND SUBJECTHOOD ACCORDING TO REGAN

Humans and animals are "subjects of a life," Regan claims, and that life has value regardless of its value or utility to any other being. His definition of subjects involves the same capacities that are catalogued in Singer's and Birch and Cobb's accounts, with the exception that Regan does not draw the "psychic" line between humans and animals that Birch and Cobb claim as the distinguishing threshold of human evolution. In many other respects, Regan's argument is very close to the process view of evolution as evolution of *consciousness*. Like many environmentalists, both religious and secular, Regan correctly presents Darwin's work as a refutation of the Cartesian split between animal and human consciousness. "Evolutionary theory," he writes,

> provides a reasonable theoretical option to Descartes' position, maintaining that consciousness is an evolved characteristic, with demonstrable adaptive value, something that is therefore reasonably viewed as being shared by members of many species in addition to the members of *Homo sapiens*.[69]

Evolution supports the conclusion that animals are conscious beings like us. They are distinct selves with "psychophysical identity" over time: their life has a *biographical*, not just a biological dimension, replete with beliefs and desires, perception, memory, and a sense of the future, including their own future.[70] Ability to experience pain is one important characteristic among others, and inclusion in the category of subjects is not dependent upon the capacity to suffer. What is immoral about our treatment of animals "isn't the pain, isn't the suffering, isn't the deprivation," Regan argues: "*The fundamental wrong is the system that allows us to view animals as our resources.*"[71] Even if animals were treated well (not allowed to suffer), many of our practices could still be condemned on the basis of a deontological rights account.

A duty not to harm subjects of a life can be derived from a principle of respect, Regan argues. Possession of inherent value implies respect for certain prima facie rights that include, minimally, the right not to be harmed, regardless of the amount, great or small, of pain inflicted or the benefits, many or few, that will result. "The respect principle rests on the postulate of inherent value" and "all those who satisfy the subjects-of-a-life criterion" have value of this kind.[72] The life of subjects can be experientially better or worse for them, and thus they can be said to have an experiential welfare. Having welfare means that it *matters* to such beings what is done to them or for them; they are recipients of both benefits and harms, "benefits consisting of opportunities for the satisfaction of desires and the fulfillment of purposes that are in the interest of these individuals, and harms being what detract from their individual welfare."[73] We fail to respect the value of such individuals if we detract from their welfare. The infliction of suffering is one example of harm that humans and animals, having a similar welfare, can be assumed to share. The death of an animal constitutes a harm as well, even if it takes place painlessly. Death, Regan argues, is a harm that occurs by deprivation rather than infliction. It constitutes the "ultimate harm" (though not necessarily the worst harm) because, of course, "death forecloses all possibilities of finding satisfactions."[74] "We have, in short, a prima facie direct duty not to harm those individuals who have an experiential welfare."[75]

Regan presents himself as a revolutionary rather than a reformer where animal ethics are concerned. Unlike Singer, he rejects even testing on animals for medical purposes that could be demonstrated as *vital*. Animals are not our resources: *any* use of them—in agriculture, sport, medicine, or labor should be abolished. Violating animals' rights is *always* wrong, Regan argues, however "extreme" such a position might seem. He points out that labeling a position extreme does not constitute a refutation of it: murder and racial discrimination are always wrong as well, and ethical consistency on these issues might also be considered extreme. "Sometimes 'extreme' positions about what is wrong are right," Regan concludes.[76] "You don't change unjust institutions by tidying them up."[77]

Are animals of all kinds, and in all types of situations, subjects of a life? The simplest answer is *yes*, but Regan does not offer much precision on this point. He declines to draw any sharp distinctions between animals with experiential capacities. Like Birch and Cobb, he points to the presence of complex "anatomical and physiological properties," such as a central nervous system, that are relevant to subjecthood.[78] At the very least, he argues, all normal mammals of one year of age or more qualify. Beyond this, the details are sketchy, but like Singer he argues for giving animals the benefit of the doubt when evidence for consciousness and subjecthood is unclear. A policy of "moral caution" is warranted, and the burden of proof rests with those who deny subjecthood. Such a policy would "have us act *as if* nonmammalian animals are conscious and are capable of experiencing pain unless a convincing case can be made to the contrary."[79] Thus we should interpret the category of subjects very broadly rather than risk harming or killing a subject of a life. The fact that we cannot draw a precise line between beings who are conscious subjects and those who are not in no way detracts from his position, Regan argues, any more than our inability to assign exact definitions to "tall" or "old" renders these categories meaningless.[80] Uncertainty about subjecthood does not warrant treating *some* animals differently from others, as we will see in Regan's discussion of speciesism.

DEGREES OF SPECIESISM

Regan rejects the argument that animals have varying degrees of inherent value: "*All who have inherent value have it equally.*"[81] Those who would exclude certain animals from having equal inherent value (on the grounds that they lack reason or sophisticated communication skills, for instance) must issue the same judgment for humans who are "similarly deficient."[82] Those who grant that *some* animals have inherent value equal to humans, while denying this status to other animals, commit speciesism as well. They are merely "modified speciesists" rather than absolute, "categorical" speciesists. In other words, a categorical speciesist denies humanlike (inherent) value to all nonhuman species. A modified speciesist believes that animals possess some inherent value (less than humans) and includes some animals (generally, those most similar to humans) within the circle of moral consideration.

Despite other similarities between their arguments, Regan would consider Birch and Cobb's criterion of richness of experience a form of modified speciesism, in that they grant inherent value only to those animals "higher up" on the richness scale. Interestingly, he would probably consider Singer's classification of animals into persons and nonpersons a version of modified speciesism as well, despite the fact that "subjects-of-a-life" and "persons" appear to be similarly defined. The primary difference is that

Singer gives higher value to *self*-consciousness than mere consciousness, while Regan speaks of consciousness in general. (Regan, perhaps, avoids being speciesist simply by being more vague than Singer about where these kinds of lines can be drawn.) In any case, Regan's condemnation of both the categorical and the modified perspectives implies that he holds *all* animals to be subjects of a life and to have the same rights.

At certain points in his discussion of modified speciesism, however, Regan begins to sound more and more like Singer. When pressed, a modified speciesist who excludes some classes of animals from moral consideration may point to a difference in something like psychological complexity in order to justify his or her position. Greater complexity implies greater value, one might argue. However, this argument cannot hold, Regan points out, due to the simple fact that "some nonhuman animals bring to their biography a degree of psychological complexity that far exceeds what is brought by some human beings."[83] Compare, for instance, a two-year-old chimp to a profoundly handicapped human of any age, Regan suggests. The argument for psychological complexity leads inevitably to the problem of "marginal" human cases. A modified speciesist will have to admit that justifications for experimenting on the chimp can equally apply to experimenting on the handicapped human. Put differently, since we would condemn experimenting on the human, we should also condemn this use of the animal. Regan and Singer agree on this point. However, whereas Regan tends to state his case in the latter form (no experiments on humans, no experiments on animals), Singer's view approaches the first form of this argument (in some instances, we could justify experiments on certain animals and certain humans). Singer remarks, for instance, "I do not believe that it could never be justifiable to experiment on a brain-damaged human." Such justification would flow from a utilitarian analysis: "If it really were possible to save several lives by an experiment that would take just one life, and there were no other way those lives could be saved, it would be right to do the experiment."[84] Singer hastens to add that such experiments would be extremely rare; furthermore, none of the experiments currently being conducted on animals fit this description, he argues.

Singer and Regan also agree that religion—particularly Judeo-Christian religion—has contributed to the mistreatment of animals and has for centuries generated (alleged) justifications for speciesism. Regan is somewhat less critical of Christianity than Singer. He suggests that, despite its record of abuse, the tradition offers some resources for a constructive ethic toward nonhuman animals. *If* Christianity could demonstrate a real spiritual difference between humans and animals (such as the idea that humans alone are created in the image of God), then this difference would be morally relevant, he argues. In fact, he believes that a morally relevant difference between humans and animals does exist, but it is not the one that speciesists generally cite. Humans are not unique among all creation except in the fact of our position of "moral responsibility" vis-à-vis other animals. "By this I mean that we are expressly chosen by God to be God's vicegerents in the day-to-day affairs of the world . . . to be as loving in our day-to-day dealings with the created order as God was in creating that order in the first place."[85] This role for humans does not lend support to speciesism of any sort. On the contrary, he argues, it offers a "healthy spiritual antidote to this virulent moral prejudice" because a proper understanding of our role should humble us and give us pause.[86] An equally powerful antidote is presented by evolutionary theory, as we have seen from Regan's argument about consciousness. Properly understood, Christianity, as well as our "best science," reinforces ethical regard for animals as our physical and psychological kin.[87]

But what about other beings who are not conscious subjects? Are insects also our kin? Trees and plants? What about animals who are conscious subjects but live in the wild—do their rights demand the same moral, loving response from us? Regan argues that his rights approach can be extended to all animals, and he condemns any approach that puts the good of the group above individual rights. Like Singer, his definition of what counts morally (individual mental and physical properties) makes it very difficult to derive an ethic for a collective entity like species and ecosystems *or* a nonsentient being such as a plant. Let us turn to Regan's discussion of species and his account of how nonsubjects such as trees might be counted morally. Following that, we will then reexamine the concept of speciesism and in light of the work of one philosopher, Mary Midgley, who criticizes animal advocates' supposedly "biological" refutation of species preference.

INDIVIDUAL RIGHTS VERSUS ENVIRONMENTAL FASCISM

Regan maintains that there is no necessary conflict between an ethic that promotes the rights of individual beings and a concern for the environment as a whole. Like Singer, he articulates a basic preference for a hands-off policy toward wild beings: "The general policy recommended by the rights view is: *let them be!*"[88] This does not mean, however, that his rights position has nothing to contribute to environmental ethics.

Any policy that takes an aggregate, such as a species, as more important than the individual rights of animals is immoral in Regan's view. For example, land ethics (in Regan's interpretation of it) proposes that "the individual may be sacrificed for the greater biotic good," and such a position, he claims, may be "fairly dubbed 'environmental fascism.'"[89] Just because something is "rare" (say, a flower), land ethics asserts that it is more important than something that is plentiful, such as a human being, Regan argues. Thus land ethics leads to the morally outrageous idea that a human being may legitimately be killed for the sake of a flower. "The rights view cannot abide this position"—not because it absolutely denies the possibility that a flower can have rights, but because it rejects the idea that "what should be done to individuals who have rights" should be decided on the basis of an "appeal to aggregate considerations."[90]

Regan proposes two options for thinking about environmental issues from a rights-based perspective. First, he argues that environmental concerns would be adequately addressed if a rights position were extended to nature, because protecting the rights of each individual being would result in preserving the whole, even though his approach does not take the whole as the important ethical unit. Rights theory remains a "live option" for environmentalists because "if we were to show the proper respect for the individuals who make up the biotic community, would not the community be preserved? And is not that what the more holistic, systems-minded environmentalists want?"[91]

This argument, however, still fails to explain how a rights-environmentalist would value individual entities that do not obviously have rights, such as plants and trees, or what Regan calls "inanimate objects." Regan's second proposal deals directly with these forms of life, which, he acknowledges, the subjects of a life concept does not easily embrace. He suggests that it may be possible (for someone) to develop a theory of rights for a tree, or collections of trees, based on their inherent value.[92] "On the rights view, assuming this could be successfully extended to inanimate objects, our general policy regarding wilderness would be precisely what the preservationists want—namely, let it be!"[93] However, Regan does not successfully develop such a rights view for trees. He merely argues that it is possible that an intelligible and persuasive case *might* eventually be made for natural objects that are not subjects of a life but nonetheless have inherent value. We should not assume that a rights view can never be extended in this direction. "Those who would work out a genuine ethic of the environment in terms of the inherent value of natural objects (trees, rivers, rocks, etc.)" should not give up in frustration owing to any features of his rights theory "since the subjects-of-a-life criterion is set forth as a sufficient, not as a necessary condition" for having value. Still, he concedes, those who attempt to locate other bases for rights "certainly have their work cut out for them."[94]

Where, then, does this leave us? Essentially, Regan's argument, like Singer's, denies value to aggregates of beings apart from the individual value that inheres in each member of that larger unit. Only individuals can have interests (in Singer's case) or rights (in Regan's). We can still value a collection of things, Regan suggests, but not in lieu of valuing each individual. Like Singer, Regan offers no convincing solution to the problem of how to value beings (or *objects*, as he calls them) that do not possess the properties of "subjects." Whereas Singer attempts to cover these remaining entities with reference to their instrumental value to sentient beings, Regan shuns all utilitarian approaches. This leaves him with a basic desire to articulate the inherent value of organisms like trees but no argument to back it up. Thus a vast number of organisms—the majority of living things on earth—are granted no moral consideration. Again, as with Singer's arguments about the environment, it seems more sensible to abandon the paradigm in question (in this case, a rights approach) as a way of understanding nature than to try somehow to make it fit where it clearly does not. Regan himself cannot successfully defend such a position of rights for all organisms, but the idea "merits continued exploration."[95] In the meantime, as they await their inclusion in our theory of rights, a vast array of species and ecosystems are disappearing.

Does the fact of endangerment make any moral difference in Regan's account? The answer is an unequivocal *no*, for the same reason it makes *all* the difference in holistic environmental ethics—namely, because only a *species* can be endangered.[96] For an ethicist such as Rolston the very idea of endangerment points to the significance of the species level and the importance of processes such as speciation: "When a biologist remarks that a breeding population of a rare species is dangerously low," Rolston inquires, "what is the danger to?"[97] It is not to the individual members, although they too will be lost, but to the form of life that the individual member represents. But, for Regan, endangered status does not entail any special consideration because a species cannot have rights, nor are the claims made by endangered species compelling enough to override prima facie individual rights. This is not to say that rights theorists are entirely indifferent to endangered species, he argues: they might still want to protect that species—or, rather, members of that species—but *not* because it is endangered. "Numbers make no difference in such a case," he argues. If we "had to choose" between "saving the last two members of an endangered species or saving another individual who belonged to a species that was plentiful but whose death would be a greater prima facie harm to that individual than the harm that death would be to the two, then the rights view requires that we save that individual."[98] Greater harm occurs in the death of a subject of a life than in the death of a being that is not a subject. If an endangered animal *happens* to be a subject of a life, it should be preserved, but not because it is endangered. Rather, in Regan's view, we would save such organisms because "the individual animals have valid claims and thus rights."[99]

Given Regan's argument that endangerment does not confer any preferential status for some organisms, it is difficult to see how respecting the rights of individual animals can lead, even indirectly, to the preservation of the *entire* community, as he maintains.[100] Respect for rights would only preserve a community of *subjects*, not the biotic community as a whole, much less an ecosystem defined as the biotic *and* abiotic components. How can these subjects even be preserved when other nonsubject species on which they depend are permitted to die (since we have no clear obligation not to harm those things)? Regan might answer that, in any case, his general policy is just to "let nature be." His response might be that if we all adhered to this policy we would avoid having to make these kinds of decisions at all. But we cannot just let it be. Regan fails to recognize that we are already implicated; we have already done *harm* because we did not respect these beings' "right" not to be harmed in the first place.

Recall that "harm," in Regan's account, entails situations where "welfare is seriously diminished" by either inflictions or deprivations.[101] He concedes that there are both natural and human causes, noting that harms of natural origin are not *wrongs* as those caused by humans are. But what are our obligations regarding past harms done—by us, not by nature? Conferral of endangered and threatened status for species—including those whose members would be subjects according to Regan's criteria—is an acknowledgment of such past harms. Leaving aside plants and other nonsentient beings, wouldn't Regan admit that, at the very least, we have a *special* obligation to protect lives of endangered subjects who are threatened by virtue of human actions?[102] Regan's interpretation of Christian obligations—as vicegerents on earth—also prescribes an obligation to protect the "created order," as he calls it. Thus his own ethics would seem to support special consideration for endangered species. But he is caught. Remedial measures, such as the Endangered Species Act, that would undo harms done to these subjects might involve inflicting *more* harm on other beings who may also be subjects.

Yet this is the reality we are faced with in environmental ethics. The situation is complex, and our ethical decisions reflect this complexity. The protection granted to endangered species—legally, since 1973 in the Endangered Species Act, as well as philosophically in approaches such as land ethics—is a recognition of our duties to protect life-forms (*forms* of life) that are threatened with extinction *because* of human actions. Protection preserves a natural process of speciation that ought not to have been disrupted by humans. "Anthropogenic extinction has nothing to do with evolutionary speciation," Rolston argues, and therefore "humans have no duties to preserve rare species from *natural* extinctions."[103] Rolston makes the same point as Regan about *kinds* of harms—i.e., anthropogenic and natural—but Regan does not seem to appreciate the implications of his own argument: natural harms are not wrong in the way that human-inflicted harms are, despite the fact that suffering is equally bad for the beings involved. "Though harmful to a species," Rolston writes, "extinction in nature is no evil in the system."[104] Rolston's comments here highlight another confusion in Regan's argument: *rareness* alone is not what generates obligations to protect endangered species in land ethics (or the ESA, for that matter) as Regan seems to believe. It is not merely *numbers* that determine our course of action; some species are naturally much rarer than others. The cause and source of harm done is crucial.

In short, Regan should be led by at least parts of his own argument to insist on protection of at least some endangered species. But he prefers not to get involved in these complicated matters, issuing instead a simplistic ethic of not harming and of letting nature be. His ethic merely ignores the backlog of harms already done and offers a vague hope that someone—not he, but some philosopher, perhaps—will devise a way of valuing the vast numbers of organisms that his ethic has failed to consider. Thus his approach has little of value to offer environmental ethics—even less, perhaps, than Singer's.

PROBLEMS WITH THE CRITIQUE OF SPECIESISM

Clearly, despite some important philosophical differences between Singer and Regan, Regan's definition of subjects of a life and Singer's category of persons do overlap. Both seek to safeguard the interests of sentient animals and both are concerned with treatment of individual animals rather than species or other collectives. Singer and Regan equally oppose the bias of speciesism, but on somewhat different grounds.

We have seen why arguments revolving around sentience, suffering, and mental level of individual animals cannot form a basis for environmental ethics as a whole, but until now we have not examined the concept of speciesism in great detail. As I have suggested throughout this discussion, environmentalists often draw on evolutionary considerations to bolster their claims about moral consideration of nonhuman animals, while ignoring other kinds of biological and ecological data that are important for contextualizing these moral claims. Opposition to speciesism is one such evolutionarily informed argument that surfaces repeatedly in environmental arguments. The basic claim is that humans and animals exist on a biological continuum and this continuum undermines preference for our own species.

Both Regan and Singer argue, for instance, that our treatment of animals (particularly animal experimentation) cannot be justified on any grounds that could not also be invoked to justify similar treatment of some classes of humans. Biological differences between humans and animals are negligible and cannot support our disregard for animal life. This, environmentalists claim, is what Darwin himself demonstrated when he argued that differences between humans and animals are a matter of degree only. Environmentalists and animal advocates are especially interested in an evolutionary continuum as evidence of the mental and physical capacities that make animals worthy of inclusion in our ethics. Regan, Birch, and Cobb (and, to some extent, Singer) argue that the process of evolution produces increasing consciousness, intelligence, and physical and psychological complexity. Therefore, they conclude, evolutionary theory forms the basis of an ethic that values animal life according to possession of *these properties*. Speciesism denies the existence of these properties in animal life.

This emphasis on the products of evolution is often combined with a *lack* of attention to evolutionary forces and ecological categories—the process of speciation, the species line that persists over time, and the suffering and death that occur among individual members of a species. Natural selection does not respect individual rights and interests; it does not protect the individual from harm, suffering, or death. As one philosopher puts it, the Society for the Prevention of Cruelty to Animals "does not set the agenda for the Sierra Club."[105] Singer and Regan's particular combination of evolutionary arguments with a prescription for protecting individual animals against harm creates an odd mismatch of data and ethics. These environmentalists want to use biologically based arguments to construct an ethic that biological processes themselves do not seem to promote or reinforce. To put the point somewhat differently, they want to do away not only with species*ism* but also with the very idea of *species* as a category that warrants moral consideration.

If we take the parallel between racism and speciesism seriously, we can see where suspicion about the importance of species might have originated: opponents of racism have discredited racial prejudice in part by pointing to the suspect nature of the concept of race.[106] From a biological perspective, there is no such thing as human "races." But the same cannot be said of species. This is the basic point that Midgley argues in her critique of the claim that speciesism constitutes a bias on par with racism.

Midgley has written with characteristic clarity on the concept of speciesism. She takes issue with the often touted similarity between racism and speciesism. Races among humans do not have biological reality—all humans can interbreed; genetic differences are insignificant, dissimilarities are superficial. For this reason, differential treatment of humans due to specious classification of "race membership" is unethical, but the same cannot be said about the concept of species. Unlike racial classifications, distinctions drawn between species are not trivial or meaningless: they are real and crucial.[107] The term *species* marks a boundary between animals that cannot (or do not) interbreed, that occupy distinct niches and are adapted to their surroundings in particular ways, morphologically and behaviorally. Rolston makes a similar point:

> The claim that there are specific forms of life historically maintained in their environments over time does not seem arbitrary or fictitious at all but, rather, as certain as anything else we believe about the empirical world, even though at times scientists revise the theories and taxa with which they map these forms.[108]

The reality of such species forms implies that it is reasonable to speak of duties to them. Midgley also understands the reality of species as a legitimate foundation for our treatment of different kinds of beings. Basing our treatment of another human upon his or her race classification is unethical,[109] whereas, in order to know how to treat an animal, it is "absolutely essential" to know what *kind* of animal it is. "Even members of quite similar and closely related species can have entirely different needs about temperature and water-supply, bedding, exercise-space, solitude, company and many other things."[110] Where animals are concerned, she argues, "to be undiscriminating is not a virtue."[111] There are differences between kinds of animals that have direct bearing on how we should treat them and what constitutes harms to them.

A related claim that opponents of speciesism make is that preference for our own species is a prejudicial bias. Again, biological data suggest that this is not necessarily the case. Indeed, preference for one's own species appears to be biologically rooted, a form of natural bonds within species. There exists a "deep emotional tendency, in us as in other creatures, to attend first to those around us who are like those who brought us up, and to take much less notice of others."[112] We will return to Midgley's argument regarding natural bonds in chapter 6. The important point to note here is that equal treatment of all animals can be unethical, whereas a preference for our own species may not be.

This does not mean that species barriers prevent members of one species from having any regard for those of another. Many species of animals show some ability to extend their social instincts and sympathies to those beyond the barrier. It is partly because of social instincts that human domestication of animals has been so successful and, at times, even beneficial to humans and animals.[113] Species bonds can, and often do, become generalized to permit consideration and sympathy for nonmembers. Presumably, Midgley might say that this process of generalizing and expanding sympathies describes what Singer and Regan are doing when they call attention to the plight of nonhuman animals. Their desire to extend concern to other species shows their humanity (if that is the right word), and the concept of speciesism has helped attract the attention of people who otherwise might never have given much thought to such issues. But the details of the argument against speciesism do not stand up to scientific scrutiny.

Midgley has hit upon a very important insight in her critique of the idea of speciesism—one that flows from more careful reflection on evolutionary considerations than the dogmatic opposition to speciesism entails. The key point clarified by her discussion, one often missed by animal advocates, is that an endorsement of moral obligations toward other animals need not imply that we have the *same* obligations toward them that we have toward humans. Differences and similarities between humans and animals are equally important. Furthermore, we may not even have the same moral obligations to all nonhuman animals. Details of obligations will depend upon several factors: the *kind* of animal in question (e.g., the species of animal, its biological needs), the context in which moral claims arise (for instance, wild and nonwild situations), and our particular relationship to the animal in question.[114] Holistic ethicists such as Rolston would add other considerations to this list: Is the species endangered or well-represented? If it is endangered, what is the cause of its endangered status? Is it a native or introduced species? But they would not ask, of organisms in the wild, *How much pain does it feel?* or *What is the mental level of this being?*

We will return to these arguments about human-animal ethics in the following chapters, as we begin to consider what our different moral obligations are to wild and domestic animals and whether or not an ethic of love provides appropriate guidelines in each of these contexts. The great value of these arguments (Midgley's as well as land ethics) lies in their demonstration that, in the rush to include nonhuman life in human ethics, important differences can easily become obscured. The desire to treat nonhuman and human life equally—the trend toward homogenization that Michael Northcott identifies—motivates many of the arguments we have looked at, both Christian and secular. A Christian argument such as McFague's that endorses a subject-subjects approach to all life is one example. Birch and Cobb treat *some* life-forms as equal to humans (those with the greatest richness of experience), as does Singer in defining what it means to be a person. Regan grants to all animals the same basic rights and the same incontrovertible, inherent value as humans. Whatever additional themes and individual strengths and weaknesses might be found in their arguments, these authors share an emphasis on locating similarities between humans and other life-forms and attributing these similarities in part to an evolutionary perspective.

The work of philosophers such as Midgley, Rolston, and Callicott affirms that evolutionary perspectives may reinforce ethical obligations toward other beings. But these perspectives also imply limits to how far and in what manner we can realistically extend certain ethical approaches beyond our own species. The desire to treat all animals as evolutionary kin must be tempered and supplemented by other considerations of what is valuable about species *aside from* values that have an anthropocentric orientation (values that benefit us directly or that stem from humanlike qualities in animals). We need to gain a better understanding of values that have a reference point outside human interests and experiences. Two perspectives—one philosophical and one theological—are of immense importance in moving environmental ethics away from its core of anthropocentrism: the ecocentric perspective provided by land ethics and the theocentric ethics of James Gustafson. The value of these two approaches for broadening the scope of environmental ethics is the topic that we turn to in the next chapter.

CHAPTER 5

Philosophical and Theological Critiques of Ecological Theology

Broadening Environmental Ethics from Ecocentric and Theocentric Perspectives

In human history, we have learned (I hope) that the conqueror role is eventually self-defeating. Why? Because it is implicit in such a role that the conqueror knows, ex cathedra, *just what makes the community clock tick, and just what and who is valuable, and what and who is worthless, in community life. It always turns out that he knows neither, and this is why his conquests eventually defeat themselves. In the biotic community, a parallel situation exists.*

—Aldo Leopold, *A Sand County Almanac*

The moral stance of the religious view I am affirming pays deeply its respect to nature, but since conflict or dissonance, as well as harmony and consonance are part of nature and our place in it, the moral stance itself does not resolve environmental ethical issues. The purposes of nature, relative to "anything that exists" and to the interdependence of all, are conflictual relative to the human good and even various "goods" of the nonhuman world. We have to ask, "Good for what?" "Good for whom?"

—James Gustafson, *A Sense of the Divine*

We understand, in spite of our wishes, that nature moves and changes and involves risks and uncertainties and that our judgments of our own actions must be made against this moving image. The message of this book is consistent with the ethical outlook . . . that "nature is not to be conquered save on her own terms." I have simply tried to give a modern view of "her" terms. It is also consistent with the land ethic of Aldo Leopold.

—Daniel Botkin, *Discordant Harmonies*

Environmentalists, whether religious or secular, agree that in order to know what our obligations are to the natural world and its inhabitants, we must acquire some basic understanding of how nature works. But how much understanding is possible? To what extent can we step outside of a human perspective and begin to envision what is good or appropriate for nonhuman forms of life? How do we decide what is most valuable in nature? How can we be certain that our ethical choices, once implemented, will have the intended effect? Environmental ethicists constantly attempt to deal with such questions, yet not all have done so successfully.

Up to this point we have examined the arguments of two philosophers and several ecological theologians whose works in environmental and animal ethics display common themes and similar deficiencies. As we have seen, the environmental perspective of many scholars in both religious and secular ethics remains inadequately informed by science and too anthropocentrically grounded. The neglect of science (or misuse thereof) and the persistence of an anthropocentric focus in these arguments are intimately related. In the present chapter, we will turn to two perspectives—one ecocentric and one theocentric—that have a number of advantages, scientifically and theologically, over those we have examined thus far.

In much of ecological theology and secular ethics, values in nature, and the ethical obligations such values generate, are articulated in language and categories that reflect human perspectives, capacities, and experiences. The assumption that individual animals in degraded environments experience a form of "oppression" fundamentally analogous to human experiences of oppression is one example; evaluative frameworks for determining the worth of nonhuman forms of life that rely on concepts of richness of experience, personhood, subjecthood, and capacities for pain and suffering likewise point to a human-centered perspective.

This is not to say that animals and humans are utterly different, physically or psychologically. Environmentalists are right to believe that Darwin's work underscores our fundamental likeness to other beings. But the important point to keep in mind is that both the similarities and the differences between humans and other animals are crucial for discerning our moral obligations. It is undeniably true that animals can and do experience pain and that many share humanlike capacities. Many can reasonably be assumed to possess a kind of rationality or subjecthood that is akin to human experience. As a basis for an environmental ethic, however, such considerations can quickly lead us astray. Thus while the interest in demonstrating similar capacities among humans and animals is in some sense laudable, many go too far in this endeavor, extending human ethics to individual animals in a wholesale and uncritical fashion. An environmental agenda motivated primarily by the recognition of similarities between human and nonhuman animals is bound to fail in terms of practical application; such an ethic also denies animals the very quality of "otherness" that many environmentalists wish to recognize. The inadequacies of ecological theology are apparent in light of scientific considerations about how evolutionary and ecological processes operate in nature, as I have argued throughout this work. However, theological and philosophical critiques expose a number of problems with ecotheology as well. An *ecocentric* perspective such as land ethics and a *theocentric* interpretation of humans, nature, and God reveal many of the same underlying problems with much of ecotheology.

In the previous chapter we looked briefly at some of the ways in which land ethicists such as Rolston and Callicott have clarified the issue of suffering in nature and the importance of preserving a larger whole, such as the species, rather than focusing on the interests, rights, and suffering of individual animals in the wild. In this chapter, and the next, we will look at these arguments in greater detail. As I hope to demonstrate, land ethics is far more consistent with natural science and provides a much more appropriate perspective from which to discern the proper role of humans in protecting natural environments. As a philosophical, rather than a theological, ethic, land ethics cannot in itself constitute a viable alternative for ecological *theology*: it does not make explicit claims about God or the relationship between God, humans, and nature. It does not (for the most part) invoke biblical passages or translate its basic tenets into theological imperatives to "love" or "liberate" nature. However, the perspective of land ethics does not necessarily *preclude* religious interpretations either. In fact, as I will argue, land ethics can be understood as consistent with a particular theological orientation. (Rolston clearly recognizes this in some of his writings in which he joins theological and biological interpretations of nature.) A nonanthropocentric theology such as that developed by James Gustafson is highly compatible with many features of land ethics. Both perspectives promote respect for forces and processes in nature, implying a more modest role for humans in caring for (rather than controlling) the natural world. Both suggest that natural processes and our ethical choices regarding them are inherently more complex and conflictual than ecotheology has realized. Land ethics and theocentric ethics strive to discern value in nature outside of anthropocentric contexts of utility or similarity to humans, and both remain consistent with scientific interpretations of nature.

The concluding chapter will be devoted to developing an alternative environmental ethic that incorporates elements of Darwinism, land ethics, theocentric ethics, and a modified ethic of love toward nature. For now, I want to begin laying the groundwork for that alternative by contrasting the perspective of land ethics and theocentric ethics, on the one hand, with those of prominent ecotheologians, on the other. First, we will examine concepts of animal suffering, sentience, and complexity (or "richness") in order to see how land ethics puts such considerations in broader, ecological perspective. We will then turn to a *theological* critique of ecological theology based on the work of James Gustafson. While not aimed at ecological theology per se, Gustafson's criticisms of Western theology and ethics expose the anthropocentrism that persists even in the work of theologians who claim to have turned their attention away from exclusively human perspectives and concerns.

We will begin with an account of the basic features of land ethics as articulated by Aldo Leopold and subsequent proponents of the land ethics approach, Callicott and Rolston. Although there are differences between Leopold's original formulation of land ethics and its current form in the work of Callicott and Rolston (not to mention differences of opinion *between* Callicott and Rolston), land ethics as a whole has been critical of anthropocentric and utilitarian interpretations of nature since its inception. Moreover, Leopold, Callicott, and Rolston all recognize the importance of taking natural processes seriously in the task of discerning what is best for biotic systems.

LAND ETHICS AND AN ECOLOGICAL CONSCIENCE

Land ethics originated in the 1940s with Aldo Leopold's classic text, *A Sand County Almanac*. In the opening sentence of this work, Leopold describes the essays contained therein as the "delights and dilemmas" of a person who cannot "live without wild things."[1] This simple sentence contains one of the most enduring insights of land ethics: its recognition that valuing and protecting wild things is an inherently rewarding as well as complex and ambiguous endeavor. *A Sand County Almanac* is the product of a lifetime of reflection on the theme of valuing wild things—a life that ended abruptly, though somewhat appropriately perhaps, when, shortly after completing the book, Leopold died of a heart attack while helping his neighbors fight a grass fire.[2] *A Sand County Almanac* was published the following year, in 1949.

Throughout his career Leopold was critical of wildlife and wildlands conservation that failed to understand nature from a holistic, systemic perspective. Conservationists, he argued, often treat natural places as though they were composed of distinct parts that can be manipulated, removed, or rearranged without causing problems for the entire system. Those whose job it was to "manage" wildlands tended to forget the most important rule of all good tinkerers, namely, to keep all the parts.[3] Wildlands managers were too willing to discard parts of nature they considered uninteresting, unprofitable, ugly, or dangerous, not realizing the importance of preserving each species in an ecosystem in order to preserve the whole. Too much management of nature (what Leopold often called "artificialized" management) created an overabundance of some organisms, such as deer, and a scarcity of others who prey upon them.[4] In the end, Leopold lamented, such shortsighted practices produced only a mass of "starved bones" of a deer herd, "dead of its own too-much."[5]

The elimination of predators in national parks was once a common management practice. "In those days, we never heard of passing up the chance to kill a wolf," Leopold recalls of his early years in forestry. He too was indoctrinated with the belief that if "fewer wolves meant more deer," then surely "no wolves would mean hunters' paradise."[6] He describes his conversion to a different set of wildlife values in terms of a radical shift from a purely anthropocentric perspective on nature to what he refers to as "thinking like a mountain": "Only the mountain has lived long enough to listen objectively to the howl of a wolf" and to find in it a meaning that is deeper than the "obvious and immediate hopes and fears" of those who briefly encounter wild nature (129). Wildlife managers who killed predators had failed to consider their value from the mountain's perspective. Similarly, the cowman who eliminates all wolves from his range does not understand that he has usurped the predator's position, Leopold writes. "He has not learned to think like a mountain" (132).

For Leopold, the perspective of the mountain symbolized a valuation of nature beyond human interest and enjoyments. He was especially critical of myopic conservation efforts that regarded nature from largely utilitarian and economic perspectives. "One basic weakness in a conservation system based wholly on economic motives is that most members of the land community have no economic value." In what "unsuspected ways," he asks, might each of these species "be essential to the maintenance" of the land? (220). Lack of economic value, he notes, is characteristic of certain species as well as "entire biotic communities: marches, bogs, dunes, and 'deserts' are examples" (212). Because we cannot determine the value of all living things—even to ourselves—a broader perspective must guide our treatment of the land.

Land ethics, he hoped, would eventually provide such a perspective. Leopold understood that purely anthropocentric justifications for preserving these larger aggregates rarely succeed. He describes two kinds of conservationists: those who regard land as a commodity with a functional value and those who perceive the value of "natural species" and prefer to "manage a natural environment rather than creating an artificial one." The latter type of conservationist values and preserves natural processes "on principle" and they alone feel "the stirrings of an ecological conscience" (221). Leopold's description of the land ethic often involves a mixture of dispositional responses (love, admiration, respect, wonder) and rule-based, principled action. These dimensions of land ethics (which I believe are fully compatible) have been developed respectively by Callicott and Rolston. Whereas Callicott often characterizes land ethics as an ethic of expanding moral sentiments, Rolston tends to speak in more deontological terms of duties and obligations to nature (though I would add that Rolston's ethic clearly expresses awe and love for nature at times). Leopold did not take a philosophical stand on the type of ethic he endorsed, except to emphasize that it was *not* purely economic or utilitarian. In the years since Leopold's death "metaethical" discussions attempt to sort out the type of moral reasoning entailed in land ethics but Leopold himself was often content to speak of "intuitive respect" or a "sense of wonder" that orients us to the proper principles regarding nature.

For Leopold, the appropriate ethic toward nature was embodied in the golden rule of land ethics, well known to all advocates of Leopoldian conservation: *An action is right when it tends to preserve the integrity, stability, and beauty of a biotic community. It is wrong when it tends otherwise* (224–25). Despite the apparent simplicity of this statement, land ethics generates specific guidelines, stemming from the basic presumption that management should be kept to a minimum and that natural forces that shaped, and continue to shape, the character of a biotic community should be permitted to operate. Preservation of natural values, with minimum intervention, is indicated by the very concept of wildness. Wilderness is something that "can shrink but not grow," Leopold argues, and thus "the creation of new wilderness in the full sense of the word is impossible" (200). In locating environmental values beyond the scope of anthropocentric values, Leopold's ethic also implies a more modest role for humans as caretakers of nature. A broader understanding of values indicates a more limited role for human action.

These features of land ethics, as I hope to show, resonate with a theocentric understanding of ethics and the emphasis it places upon values beyond human values. However, we have yet to consider another important feature of Leopold's argument, namely, the "implicit framework of ecological and evolutionary theory" that underlies the land ethic.[7]

DARWINISM AND LAND ETHICS

"I have purposely presented the land ethic as a product of social evolution," Leopold writes, "because nothing so important as an ethic is ever 'written.'"[8] What did Leopold mean when he spoke of land ethics as an evolutionary possibility? Following Darwin, Leopold describes an evolution of moral consideration that begins with basic social instincts toward kin and gradually expands to include all life. In its initial stages, ethics is limited to one's own family or immediate social group (rooted in bonds between parents and offspring and among immediate kin). Eventually, ethics includes not only all members of one's species but ultimately extends to all living things, including the land itself. This is the basic scheme of moral evolution described by Darwin in *The Descent of Man* (1871). Leopold, however, takes moral evolution a step beyond Darwin's account, extending moral consideration not only to animals but to the land itself. "The land ethic simply enlarges the boundaries of the community to include soils, waters, plants, and animals, or collectively: the land."[9]

According to this view (further developed by Callicott), moral sympathies have become more and more expansive over the course of evolution, and we now find ourselves considering, perhaps for the first time in history, the possibility that an entire aggregate of land can be part of our moral community. The "logic" of this position, Callicott argues, "is that natural selection has endowed human beings with an affective moral response to perceived bonds of kinship and community membership and identity." One's concept of community would come to embrace the "ecological community" as well. Callicott argues that the "conceptual foundations" of land ethics can be traced in part to a "Darwinian protosociobiological natural history of ethics, [and] Darwinian ties of kinship among all forms of life."[10] The development of a land ethic makes sense in light of Darwin's account of the evolution of ethics.

We can see, then, that like many environmentalists, Leopold also cited the Darwinian discovery of biological continuity as a catalyst for developing an ecological consciousness. In the wake of evolutionary theory, he argues, we now know something that previous generations did not understand about nature. Darwinian knowledge "should have given us, by this time, a sense of kinship with fellow-creatures; a wish to live and let live; a sense of wonder over the magnitude and duration of the biotic enterprise."[11] Above all, Leopold believed, Darwinism has provided the crucial insight that while human beings may be considered the captain of the "adventuring ship" called Earth, we are "hardly the sole object of its quest."[12] A scientific perspective broadens our understanding of life and the human place within it.

At first glance this Darwinian/Leopoldian interpretation of environmental ethics appears to agree, point for point, with the arguments of other environmentalists we have examined. Like Singer and Regan, Leopold suggests that we simply extend our existing moral framework to include all other forms of life. At times he even argues that members of the biotic community have a biotic "right" to exist.[13] His understanding of the need to include all species, and the land itself, as part of our moral "community" seems to concur with ecotheology's account of the ecological model, in which our ethical obligations derive from a sense of community with our evolutionary kin. Like land ethics, the ecological model emphasizes an expanded network of moral relationships. In *A Sand County Almanac*, as well as other essays, Leopold even castigates the "mechanization" of nature and denounces a purely utilitarian evaluation of the natural world, much as ecotheologians do in repeatedly urging us to reject mechanistic, objectifying attitudes toward life. Humans "cannot destroy the earth with moral impunity," he argues, "because what we think of as 'dead' earth is an organism possessing a certain kind and degree of life, which we intuitively respect as such."[14] Leopold, Darwin, and ecotheologians appear to agree that a proper ethic toward nature entails a recognition of all life-forms as our kin, deserving of similar treatment as members of our moral community.

But there are important differences. As Callicott notes, land ethics respects and preserves fundamental "trophic asymmetries," such as those between predator and prey, that lie at the very heart of ecological communities, regardless of how unjust or cruel these may seem from a human perspective. For instance, Leopold's model of nature embodies the sort of hierarchical relationships—trophic levels—that ecotheologians banish from the ecological model. "The image commonly employed in conservation education is the 'balance of nature,'" Leopold notes, but "this figure of speech fails to describe accurately what little we know about the land mechanism." He proposes instead a more appropriate image, "the biotic pyramid."[15] The entire pyramid functions by means of a process of eating and being eaten in turn, he argues. Each "successive layer depends on those below it for food . . . and in turn furnishes food and services to those above."[16] Interdependence is a central feature of this model, as it is for the ecological model, but Leopold stresses that ecological relationships are comprised of predator-prey and parasitic interactions.[17] "Thus," Leopold writes, "for every carnivore there are hundreds of prey, thousands of their prey, millions of insects, uncountable plants." Individual organisms live and die continually, but the species line continues. There is a certain stability to this structure but not harmony, for the pyramid is "a tangle of chains so complex as to seem disorderly."[18]

Leopold's pyramid image suggest some interesting things about the processes of nature and the attempt to derive moral implications from ecological "communities." Human concepts of justice, mutuality, and equality do not find affirmation in such a model. Perhaps for this reason, some ecotheologians attempting to derive norms from nature downplay the role of natural "disvalues" such as predation and death. In fact, some Christian environmentalists are quite explicit in excluding such interactions from consideration: In *Earth Community, Earth Ethics* Larry Rasmussen observes that nature is "too casual" about suffering for humans to look to it for clear norms. Predation, though part of nature, is "not a pattern of morality we praise and advocate." Violent features of nature are thus "rejected here as a moral paradigm for sustainable community," even while nature's more benignly communal dimensions are celebrated and adopted as normative.[19]

Land ethics, by contrast, respects as good and endeavors to protect the "very inequities in nature whose social counterparts in human communities are condemned as bad and would be eradicated by the familiar social ethics," Callicott writes.[20] Moral consideration of the *individual* member of the ecological community is "preempted by concern for the preservation of the integrity, stability, and beauty of the biotic community."[21] The claim that the preservation of the whole can trump the "rights" of the individual is, of course, precisely what Regan sees as fascistic about land ethics. Land ethics is "holistic with a vengeance," as Callicott admits. But an ecocentric ethic demands that we value the processes that generate species, even when this process does not suit human moral preferences. "One cannot properly value wildlife without valuing the forces, ancient and ongoing, that create wilderness"; conservation requires a longer-range perspective than most human endeavors. "Only those able to see the pageant of evolution can be expected to value its theater, the wilderness," Leopold writes.

What does Leopold mean by the "biotic community"? His use of this term takes a "middle route" between cooperative, superorganismic concepts, on the one hand, and nature as a fortuitous, conflict-ridden, "accidental jumble" on the other.[22] In this sense, Leopold's understanding of ecology appears quite modern. When he claims that we should include the land as part of our moral community he does not mean that the ecosystem is a single organism whose health and life must be safeguarded as we would safeguard a human life; nor are we to be solicitous of each individual member of biotic communities, caring for, feeding, and healing them as we would a fellow human in need. To do so would destabilize the land pyramid, whose "lines of dependency" are constructed out of complex "food chains."[23] The hierarchical structure of the system is sustained by the deaths of its individual members. "A fauna and flora, by this very process of perpetual battle within and among species, achieve collective immortality."[24] The ecological community and *human* communities will not necessarily point to the same set of ethics. Cooperation in the biotic community is not intentional in the same sense that human cooperation is, and the "community" as a whole neither oversees the well-being of its members nor ensures in some teleological sense that all needs will be gratified. As Rolston observes, "Plants do not intend to help falcons and cheetahs, nor does any ecosystemic program direct this coaction."[25] But while the biotic community and its members do not act as moral agents, do not abide by a set of ethics, they are worthy of moral consideration, as well as places of "value capture."[26]

For Leopold, the Darwinian idea of kinship with animals (which he clearly admires) does not imply that we are literally to treat them as our family or "neighbors." In the passage from *A Sand County Almanac* in which Leopold rehearses the lessons of Darwinism, he highlights other insights, in addition to kinship with nature: a "sense of wonder" over the "magnitude and duration" of the *whole biotic enterprise* and the realization that humans are not the "point" of nature. The sense of wonder is increased—or ought to be—by the recognition that, in Leopold's words, we are not the "sole object" of life's quest. All these implications are important; biological continuity alone does not capture the full import of Darwin's message. In short, the ecological community cannot mean the same thing in land ethics as it means in ecological theology, nor can individual organisms have the sort of rights and interests that Regan and Singer ascribe to them. Continued existence of a larger whole can be promoted only by respecting natural processes *on principle*, regardless of the cost to individual beings.

For all his eloquent reflections on the value of wild places and the evolution of an ecological conscience, Leopold was not a philosopher, and he did not work out the precise philosophical underpinnings of the land ethic, nor could he have foreseen developments in ecology that would alter the way in which concepts such as stability are understood (more on this in the section on ecosystem health and stability according to land ethics). He argued that the values embraced in conservation must be "broader" than those usually understood by humans. "I mean value," he stressed, "in the philosophical sense."[27] Further development of Leopold's philosophical sense of value and refinements of his ecological concepts have fallen to current generations of land ethicists and holistic environmentalists such Callicott and Rolston.[28] These philosophers have carried on his work with certain modifications, in light of developments in ecology and philosophy since Leopold's time.

LAND ETHICS SINCE *SAND COUNTY*

Although Callicott and Rolston trace elements of their environmental philosophies to Leopold's holism (and, as will see, to Darwinian thinking as well), they sometimes disagree on fundamental issues, such as what is, or ought to be, designated by the term *wilderness* and what it means—in a philosophical sense—for nature to have value.[29] At the heart of their disagreements is the question whether intrinsic values in nature are human-ascribed or arise independently of humans—that is, whether or not they are really "in" nature at all. Rolston takes the position that they are; Callicott disagrees.[30] We will examine Rolston's view of natural values more closely when we turn to a discussion of "richness" and "complexity" in the natural world.[31] For now it is worth emphasizing that Rolston's view of (nonanthropogenic) values incorporates "disvalues" as well—e.g., pain, death, disorder in nature. Moreover, he wants to maintain that, had humans *never evolved* to judge values and disvalues in nature, they are in fact there.[32] Nevertheless, his commitment to a nonanthropogenic account of disvalues such as pain does not lead Rolston to promote an ethic that *eradicates* pain.[33] In my view Rolston's refusal to align himself with an ethic aimed at the prevention or removal of natural suffering distinguishes him from process accounts such as Birch and Cobb's, which similarly acknowledge the existence (and in some sense even the *necessity*) of suffering while defending an ethic that would eradicate it. Put differently, Rolston appears to agree that natural values and disvalues (what process thinkers might refer to as enjoyment and discord in life) are necessarily intertwined, a "package deal," as one theologian puts it.[34] But Rolston's ethic endeavors to respect this form of interdependence rather than censuring it. Moreover, he does not urge prevention of suffering, and promotion of well-being, for each *individual* organism, arguing instead that disvalues at the level of individual interactions (such as predation) can contribute to values at other levels, over longer periods of time than the individual's lifespan.[35] As Robin Attfield argues, in a system of values and disvalues as envisioned by Rolston "nature does not observe social justice or compensate individual sufferers, but the overall system is not one of unalleviated misery."[36] In any case, while Rolston and Callicott disagree about the anthropogenic sources of value and disvalue in nature, they agree that natural communities should not be subject to censure from the standpoint of human morals and communities.

As far as the term *wilderness* is concerned, Callicott maintains that a sharp distinction between human culture and wilderness reflects an unscientific bias—a "pre-Darwinian Western metaphysical dichotomy between 'man' and 'nature'"—that fosters a naive belief that nature, until recently, remained stable, static, and largely unmodified by human habitation."[37] Rolston argues for a clearer separation of human culture and its products from nature, and charges Callicott with denying any distinct *intrinsic* value to natural places insofar as he blurs cultural and natural values.[38] The human species, Rolston believes, now exists in a more or less postevolutionary phase in the sense that pressures of natural selection no longer have a direct impact on us: "Humans help each other out compassionately with medicine, charity, affirmative action, or head-start programs."[39] There is nothing unscientific, he argues, "about the claim that innovations in human culture make it radically different from wild nature."[40] But, again, while they disagree on important conceptual issues, ethicists such as Callicott and Rolston highlight precisely what is wrong with much of contemporary environmental ethics, namely, its focus on human-centered values and modes of experience that obscure the importance of other values.

NATURAL VALUES VERSUS ANTHROPOCENTRIC VALUES

Despite rapid developments in environmental and animals ethics in the last thirty years, Rolston laments that there has been no real "paradigm shift" in our attitudes toward nonhuman animals. Echoing Leopold's criticisms from half a century ago, he argues that we still have not managed to articulate ethical obligations toward nature from a nonanthropocentric and nonutilitarian perspective. "We do not yet have a biologically based ethics," so long as we remain preoccupied with a "psychologically based ethic" that gives priority to human kinds of experience.[41]

We have seen that Rolston argues for the significance of species—and *processes* of speciation—as valuable, and we contrasted his views with those of Regan and Singer. In what follows, we will again turn to a few cases in environmental ethics that illustrate why the arguments of ecotheologians likewise fail to shift the focus of ethics beyond narrow anthropocentric concerns. The considerations that guide ecotheology are not drawn from a study of nature but are merely a human set of concerns and interests grafted onto nature. A closer look at these cases also reveals the multitude of environmental factors that have to be considered and weighed in a land ethics approach. Decisions about what is "best" for nature inevitably involve conflicts and trade-offs that ecotheology does not take seriously enough. Interpreting ecological communities as similar to (or as an ideal model of) human communities obscures the fact that environmental ethics cannot ensure the well-being of each individual member of the community, regardless of those beings' degree of sentience or mental sophistication. Furthermore, in articulating values beyond anthropocentric ones, an ecocentric ethic such as land ethics is better able to preserve species who possess none of the capacities that environmentalists often cite as valuable. Finally, these cases also clarify why an ethic that is cautious about intervening in natural processes does not simply default to a policy of "letting nature be," as Singer and Regan suggest. An appropriate ethic toward nature demands that we take action, even when the actions that are indicated result in unfortunate outcomes, even death, for some of the individual beings involved. Environmental ethics requires us to make exceedingly difficult choices.

Real-life cases in environmental ethics demonstrate the context in which a land ethics approach interprets and prioritizes conflicting values. The following examples are taken from Rolston's works, *Conserving Natural Value* and *Environmental Ethics*, as well as from the National Research Council's exhaustive study, *Science and the Endangered Species Act*. Many of Rolston's examples deal with ethical dilemmas faced by managers in national parks whose policies adhere to holistic and land ethics approaches.

In Yellowstone National Park, in 1983, a bison fell through the ice in the Yellowstone River. A group of snowmobilers attempted, against park officials' express wishes, to pull the struggling bison out of the ice. The snowmobilers pointed out that a drowning human would have been rescued immediately, but the park officials refused to rescue the bison. One snowmobiler demanded that the park at least shoot the suffering bison if they were not going to save it, but the park's ethical guidelines (which generally follow a land ethics policy) indicated that the animal should be allowed to suffer and die, without intercession from the park.

In a second case in Glacier National Park, a few years later, a wolverine attacked but did not kill a deer. The deer suffered through the rest of the day and night and finally died the next morning. Again, park officials were criticized for what seemed an inordinate respect for the cruelty of nature. Their actions seemed even more incomprehensible when, later that spring, a bear hit by a truck was killed by park officials so that it would not have to suffer.

Yet another case that attracted a great deal of attention from the media, as well as animal rights groups, involved killing entire herds of goats in order to preserve three species of plants. In San Clemente Island the U.S. Fish and Wildlife Service and the Navy ignored the protests of animal groups such as the Fund for Animals and shot fifteen thousand goats (as many as fourteen thousand were also removed from the island with the help of the Fund for Animals).

Populations of certain marine mammals such as sea lions have risen sharply in recent years as a result of the establishment of the Marine Mammal Protection Act. The increase in populations of sea lions has led to overpredation of fish species such as steelhead, a situation exacerbated by steelhead's tendency to congregate in human-made locks, where they are more easily captured and eaten by sea lions. In 1994 the Marine Mammal Protection Act was amended to permit killing marine mammals under certain conditions, and the state of Washington proposes removal and, if necessary, killing of sea lions to prevent them from preying on steelhead.

Is there a consistent ethical argument that connects all these cases? Clearly, from the perspective of an ethic that valued animals according to their richness of experience or ability to suffer, the judgment of park officials seems schizophrenic. In one case a bear is mercifully put out of its misery, but the same consideration is denied the suffering deer and bison. Perhaps wildlife managers consider the bear's richness of experience to far exceed the deer's or the bison's and therefore acted to reduce its suffering. But if this were the rationale, why would goats be killed in order to save plant species that are nonsentient and have no capacities for experience? Why would an intelligent and complex species of marine mammals be culled for the sake of a far less sophisticated fish species?

Obviously, other factors guide managers' decisions. The deer and the bison suffered, and ultimately died, of causes that the parks deem "natural." Furthermore, these animals were not endangered, so their individual deaths would not perpetuate the decline of the entire species. The bear, on the other hand, was injured by causes that were not natural: "The encounter with a truck (an artifact) is not part of the forces of natural selection," Rolston explains. "Where humans cause the pain, they are therefore under obligation to minimize it."[42]

Compassion (and actions motivated by compassion) are warranted in some cases. In another situation involving bear species, a female grizzly and her three cubs crossed the ice on Yellowstone Lake and reached an island where they found, and began to consume, two elk carcasses. Days later, the ice melted and the bear and her cubs were stranded on the island without enough food to sustain them (the mother could swim the distance but would not leave behind her cubs who could not do so). Park officials rescued all four bears. Why? The cause of the bears' suffering was natural, so their intervention cannot be justified on grounds that an artifact of culture, or other anthropogenic causes, imperiled them. It might seem that park guidelines allow nature to take its course in an inconsistent fashion. However, because these individual bears belonged to an endangered species they were given special consideration from park managers. "They were not rescuing individual bears so much as saving the species"[43]

The fact that bears are a sentient and—at least for many people—a fascinating animal is beside the point (although, as we will see in a moment, holistic environmentalists do not by any means dismiss the special attraction that certain large, beautiful, or fierce animals hold for many people). In this case the bears are more like the endangered but insentient plants than they are like the goats. In fact, the goats differ significantly from the bears in that the former are both a plentiful and an *exotic* species, that is, a feral species introduced into the ecosystem by humans. Therefore, several issues have to be considered before deciding what to do in each case: is the cause of the problem natural or was it inflicted in some way by humans? Is the species native or exotic? Is the species endangered? Is its endangerment the result of human actions? In the case of the rescued grizzlies, endangerment is not the sole issue; also crucial is the fact that "humans had already and elsewhere imperiled the grizzly" (a point that is lost on Regan).[44] Our past actions dictate a current duty to act—to intervene—so as to preserve the species and try to remedy previous errors. "Some human interventions are more, others less natural, depending on the degree to which they fit in with, mimic, or restore spontaneous nature," Rolston argues.[45] So letting nature take its course does not simply amount to adopting a hands-off approach in all situations, giving nature a "blank check."[46]

Yet past events also constrain our ability to intervene in ways that will not cause suffering to any of the beings involved: in the case of the goats killed to save a plant species, the introduction of these animals by humans has contributed to the problem. Here the issue becomes especially complex because it is not the goats' "fault" that they were brought to live in their present habitat. Killing them seems particularly unfair, since the reasons for their overrepresentation and consequent harm to the ecosystem can be traced to previous human intervention. The same point holds for the sea lions, whose population explosion (and ready access to steelhead in northwest locks) resulted at least in part from human actions and errors in judgment. But a difficult choice has to be made, and land ethics would favor returning the ecosystem to a more natural state and protecting a species endangered by the goats' or sea lions' presence, even though this involves the death of sentient beings.[47]

With these cases in mind, let us now consider some of the values that ecotheologians (as well as some secular philosophers) have put forth as the source of ethical obligations toward nature and see how these values differ from those that land ethicists seek to promote. As it turns out, land ethics and ecotheology recognize and seek to preserve some of the *same qualities* in animals and nature. Indeed, it is important to note that the *basic intuitions* about what is valuable, according to ecotheologians, are not necessarily misguided.[48] The belief that health in nature has meaning and is a good thing, the humane conviction that suffering is in some sense wrong or tragic, and the acknowledgment that qualities of animal life—complexity, mystery, beauty, intelligence—genuinely have value, all reflect a perspective that is commendable.

This general orientation of concern, admiration, perhaps even love toward nature is in itself a positive impulse. The problem lies in the perspective that seeks to understand nonhuman suffering and sentience as essentially similar to humans' and posits a role for humans in nature that is too controlling. I want to suggest that it is entirely possible that at least some of the basic intuitions of ecotheology can be reconciled with a more appropriate, ecocentric perspective. This perspective would allow for the expression of many of these concerns and values and yet limit the scope of their application. Again, a broader, less anthropocentric perspective entails a more restricted role for human intervention in natural processes. Land ethics is not necessarily indifferent to the concerns about nature and animals expressed in ecotheology but seeks to temper these with respect to 1. the natural context in which they arise and 2. real differences between human and nonhuman forms of life. We turn now to a discussion of a few of these values, including richness, complexity, and diversity of life-forms as well as concepts of natural health, sickness, and suffering.[49]

COMPLEXITY

The complexity of certain forms of life is often seen as a reason for valuing and preserving them. Process theology, for example, holds that the processes of evolution produce ever increasing complexity of life; there are "emergent" values in evolution. Humans display the greatest evolutionary complexity, but many other species are complex as well and have considerable value. Land ethics would agree that evolution has produced some forms of life (including humans) that are more complex than others, but maintains that complexity as an *isolated* value makes little sense when deciding whether and how to intervene in nature. Complexity in nature must be supplemented by other natural values such as diversity.

Recall that Birch and Cobb argue that our obligations are the greatest toward life-forms that demonstrate the greatest richness of experience. Land ethics suggests that management with an eye toward promoting (or otherwise giving preference to) values of complexity that Birch and Cobb endorse could lead, ironically, to a loss of diversity and complexity at the level of an ecosystem. In valuing certain forms of life over others (as when conflicts emerge between an organism with a high degree of inherent value and one with primarily instrumental value), preserving richness and complexity at the expense of the proliferation of the "lower" species decreases *overall* richness and complexity. Ecosystems require a diversity of "functional roles, for example, autotrophs and heterotrophs, predators and prey"—regardless of their degree of complexity. Valuing complex forms, apart from their environmental roles and relationships would lead to a simplification of biotic life. "There cannot be higher forms all by themselves."[50]

Is it always wrong, from the perspective of land ethics, to want to emphasize or protect complex, beautiful, or remarkable organisms? Despite a holistic orientation, Rolston is not at all contemptuous of the intuition many people feel that we have a particular obligation toward certain "charismatic" species. He acknowledges that some animals seem to us "especially gifted—

the whales and dolphins, or the big cats, or the primates" (63). Even though some conservationists might initially balk at his argument, he maintains that, if we had to choose between saving a species of beetles and one of gorillas, all things being equal, we should probably save the gorilla. At first glance, Rolston's special consideration for charismatic species sounds similar to Birch and Cobb's position; it seems to be in tension with other values such as biodiversity and the value inherent in *any* process of speciation, irrespective of its products. In fact, one might even say (and some critics have said) that Rolston's understanding of values remains hierarchical, that he gives priority to vertebrates, mammals, and primates.

But preference for such species must still be supplemented by other natural and systemic principles—principles that can override concern for charismatic species. All things are *not* equal in nature, and the choice between a charismatic species and a noncharismatic one can be complicated. For instance, in the case cited above where goats are pitted against an endangered plant species, Rolston concedes that goats have certain charismatic properties as one of "the most nimble and sure-footed creatures on earth" (65). But a cluster of ecological considerations mitigate these grounds for preserving the goats, as we have seen. Chief among these is that if the goats are not culled the species of plant will become extinct, whereas eradicating individual goats will not mean the end of *that* species. The conflict therefore is not an all-things-being-equal scenario: there are other factors to be weighed. "If the tradeoff were merely one on one, a goat versus a plant, the charismatic goats would override the plants," Rolston explains. But conflict is, rather, between *individual* goats and whole *species* of plants. "The well-being of plants at the species level outweighs the welfare of the goats at the individual level" (65). Charisma is not necessarily the dominant value and does not in itself determine our course of action. Even where charisma is a more dominant value, it is not clear that a "hierarchy" that might emerge from such considerations (more charismatic, less charismatic) would be an *anthropocentric* hierarchy, for the very reason that charisma, as Rolston defines it, embodies an admiration of nonhuman animals for their distinctly nonhuman qualities. A proper appreciation of charismatic animals stems from a recognition of a grace and skill that is unlike that of humans, qualities that border on the sublime in their ability to evoke fascination and perhaps even fear in us.[51]

In commenting upon past wildlife management errors, the National Research Council notes that "management that views each species as an entity by itself, with no or little attention given to the network of interactions, is likely to produce faulty protection strategies."[52] These kinds of ecological considerations suggest that we cannot "confidently affirm" with Birch and Cobb that because porpoises, for example, possess a more "rich subjective life" they make "claims upon us beyond those made by tuna and sharks."[53] Our inclination to preserve extraordinary species should not simply stem from a recognition that porpoises are animals (relatively) "like us" and therefore deserve the sort of moral consideration and respect that we do. "The animal excellences are not valuable as poor imitations of what is later achieved in humans," Rolston argues.[54] Yet the process account of evolution of consciousness assumes an ascending order of consciousness and complexity, with humans as the end product of a process that has distributed more or less similar properties among other forms of life as well. Moral extension based upon animals' similarity to humans is inadequate and produces an ethic that is both anthropic and subjective.[55]

RICHNESS

The concept of richness is closely connected to complexity and is often cited as a value in human and nonhuman life. For Birch and Cobb, richness means experience resulting from physical complexity (possession of a large brain, a sophisticated nervous system as in the "higher vertebrates"). Richness of experience generally *corresponds* to complexity, but is not exactly the same thing because the former points to quality of life issues (i.e., experiential issues) and not merely to quantifiable physical characteristics. Rolston would agree that we can locate something like "richness" in animal life, but, again, the problem we encounter in emphasizing such traits is that most such theories of value are too "human-centered."[56] Broader, ecological considerations illustrate how a richness criterion often fails to adequately address the larger context in which such values emerge.

For instance, when Rolston speaks of "richness" he has in mind the sort of "systemic richness"[57] that does not inhere merely in individual beings or even in a species as a whole. Systemic richness refers to the processes in nature that occur among living things, such as "competition between organisms . . . plant and animal successions, speciation over historical time"—forces that "generate an ever-richer community."[58] A systemic interpretation of richness locates it in "the productive process" itself.[59] On this account, it is clear why we cannot have true richness in a zoo, even if it contained a wide array of animals, all of them possessing their own *individual* richness or complexity (say, if all

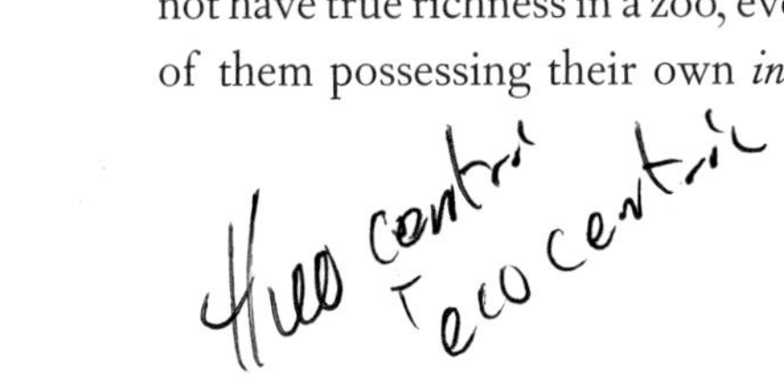

rated high on Birch and Cobb's scale of animal life). Systemic richness is found in the "evolutionary toil, elaborating and diversifying the biota."[60] Leopold alludes to similar values when he suggests that it is the whole biotic enterprise that must be appreciated and respected. Systemic values are reflected by the biotic community as a whole; they are values projected outward into nature, yet many ethicists are drawn to the "duties-to-subjects-only position" because these are duties with which we are already familiar.[61] Animal rights ethicists such as Tom Regan assume that in fulfilling ethical obligations to each individual within a group we will, by extension, fulfill those of the group as a whole. Subjectivist ethics of this sort fail to appreciate the richness that belongs to the system as a whole, not merely its parts.

Subjectivist definitions of richness might also lead people to believe that plants, for instance, do not possess value (Birch and Cobb, as we have seen, ascribe primarily "instrumental" value to plants precisely because they lack individual complexity and therefore richness of experience). A more appropriate valuing of nature's richness requires breaking out of "psychological restrictions on felt experience and to see that a forest full of plants, shrubs, herbs, that are flourishing does constitute a richness."[62] Indeed, the bulk of the earth's biotic richness is found in such organisms, since plants, insects, and other "lower" life-forms make up a much larger portion of the earth's biomass than do the large, complex vertebrate animals. Clearly, "we need language here that is not exclusively oriented to psychological states," Rolston concludes, "much less human experiences."[63] This understanding of richness would not exclude certain organisms, such as the unlovable but endangered mussel species discussed in chapter 3.

It is particularly ironic that Birch and Cobb fail adequately to appreciate this sort of systemic, interactive richness in light of their insistence on the ubiquity of *process* in nature and their preference for "event thinking" over "substance thinking" (where one grasps the primacy of "internal relation" rather than treating organisms as isolated parts).[64] Event thinking, in other words, embraces the relational, ecological model rather than atomistic, mechanistic ones, Birch and Cobb maintain. The ecological model is "based on internal relations. It asserts that living organisms take account of their environments, that is, that their relations to their environments are constitutive of what they are" (105). Yet their understanding of richness and complexity is often at odds with systemic concepts.

At times Birch and Cobb seem to perceive some of these difficulties with their approach. At one point they assert that the "right of the species to survival is much more fundamental than the right of the individual member to life" (170). Some consideration, they acknowledge, ought to be given to the "web of life" and the "value of diversity" when seeking to maximize richness of experience. *How* such issues are to be weighed and compared (richness as opposed to diversity, species versus the individual) is never clarified. Their occasional use of "rights" language further clouds the issue, since Birch and Cobb do not consistently defend a rights position. The basic task before us, they argue, is "identifying appropriate rights for different types of animals," with the assumption that not all animals have the "same rights as humans." Doing so "requires that we consider carefully what characteristics of living things give them particular claims upon us and that we make some rough estimate as to which species have the characteristic in question" (175). But, as Birch and Cobb have already established, those animals possessing the greater richness have the greater claim upon us—they have the "characteristic in question."

The priority of richness over other considerations is also clear in Birch and Cobb's argument that (nonhuman) animals possessing a high degree of richness cannot rightly be treated as mere means, whereas plants can. For such animals "existence and enjoyment is important regardless of the consequences for us or for other entities" (153). How important is their existence and enjoyment? Clearly, less important than that of humans ("emphasis is put upon the quality of human life"; 173) but more important than those beings with less richness ("porpoises as a whole are of much more intrinsic worth than sharks"; 155). Thus considerations of overall diversity remain secondary at best to richness in organisms. "In proportion to their capacity for rich experience we should respect them and give consideration to making this experience possible" (153).

In my view Birch and Cobb's richness criterion fails to capture what is most significant about Rolston's account of charismatic species. Our response to these organisms flows from a recognition of grace and skill, "competence and virtuosity," often completely alien to us. Compare this interpretation of value to Birch and Cobb's commitment to locating similarities between humans and animals as a rationale for valuing "life." "The brain," they suggest, "is extremely important for experience and similarities between human experience and that of other organisms with developed brains is likely to be much greater than similarities with organisms not so equipped." But, they hasten to add, we should not despair of finding any grounds for valuing animals different from us, because "what is remarkable in light of this enormous difference are the similarities still to be found" (125). Rolston, in other words, understands dolphins, big cats, and other charismatic animals as especially valuable not because they most closely approximate human richness but because they possess skills, talents, ways of existing in the world that are mysterious and foreign. In order to appreciate this, we have to understand what the word *charisma* alludes to, he advises. The term (*charis*) means "grace," conveying a sense of "giftedness" and uniqueness at which we can only wonder. Rolston's interpretation of the special value of charismatic animals already contains within it an appreciation of otherness, difference, and diversity; Birch and Cobb's account of richness as *similarity* does not.

Nevertheless, critics may charge that Rolston's ethical framework remains hierarchical and that, furthermore, the hierarchy he endorses retains some ties to concepts of sentience and self-consciousness. If so, he hardly seems to have broken with conventional rankings of "higher" and "lower" organisms in nature, with humans at the top and nonsentient, nonself-conscious beings at the bottom. Mark Wynn notes, for example, that some radical environmentalists and deep ecologists may regard Rolston as having "conceded too much" to anthropocentrism in "allowing that [quoting Rolston] 'the highest value attained in the system is lofty individuality with its subjectivity, present in vertebrates, mammals, primates, and preeminently in persons.'"[65] At the same time, others may object on quite opposite grounds that Rolston is too ecocentric, too affirming of this world and its processes, and that he thus fails to "take seriously enough the doctrine of the Fall," foreclosing the "possibility that the natural world will be subject to radical improvement at the time of the eschaton."

Given Rolston's call for a paradigm shift in environmental ethics—i.e., his insistence on a value framework that clearly eschews the primacy of sentience and psychological criteria,[66] his nonanthropocentric definition of charismatic animals discussed above, his rejection of "duties-to-subjects-only" positions, and numerous other ecocentric arguments—the first criticism misconstrues Rolston's work. Immediately following the sentence quoted by Wynn regarding lofty individualism, Rolston argues that "the most valuable of the parts is of less value than the whole."[67] Just as his recognition that real suffering exists in nature does not generate a mandate to remove it, so his recognition that there is "concentrated value" in certain complex, individual organisms does not translate into an ethic that gives preference to those individuals a priori.

What about the second charge against Rolston, namely, that his emphatically earth-bound orientation is potentially at odds with Christian expectations of a restored and perfected creation? In my view Rolston's rejection of redemptive, eschatological improvements to nature is one of the chief strengths of his position, both scientifically and theologically. With this potential criticism of Rolston in mind, let us examine more closely the different way in which land ethics and ecotheology understand concepts of suffering, health, and redemption in nature.

SUFFERING, HEALTH, AND REDEMPTIVE ENVIRONMENTALISM

Ecotheologians need to be wary of interpreting natural "health" or "well-being" in terms of a biblical (original, uncorrupted) state, the reestablishment of which is often presumed to be the overriding goal of environmental ethics. Discussion of concepts of health and suffering in ecotheology leads inevitably to theological assumptions and hopes about nature's genesis and its *ultimate* fate. The suffering of nonhuman forms of life is a concern that takes different forms and is expressed with varying degrees of emphasis in the authors we have looked at thus far, but there is a general expectation in this literature that humans are obligated to promote, protect, or restore a more healthful condition to nature. As with richness and complexity, there is nothing necessarily wrong with the desire to see nature preserved in a state that we can reasonably refer to as healthy. There may be a legitimate sense in which environments, and the species that inhabit them, have become sick or diseased; it is understandable that environmental ethics responds to suffering that ought not to occur. But these concepts, too, must be put in their proper perspective, with greater critical attention given to the context as well as the assumed role of humans as nature's healers.

Concern for suffering is paramount in much of ecotheological literature. In the account of environmental value offered by Birch and Cobb, for instance, richness of experience and concern for suffering are related: animals having the greatest richness of experience can be assumed to suffer the most, and such animals make the greatest moral claim upon us. Sallie McFague, too, regards animal suffering as a form of human oppression and urges Christians to extend an ethic of love to this oppressed class. Rosemary Ruether promotes an ethic of earth "healing" (despite the fact that her definition of ecology assumes the existence of a "healthy and life-giving balance" in nature) and, like McFague, she identifies certain parallels between the suffering and exploitation of nature and that of women and other oppressed classes of humans. The eschatological theology of Jürgen Moltmann also looks to a time when all struggles and suffering in nature cease, when peace, health, and wholeness are restored. Often in these accounts our role as God's "proxy" or "representative on earth" suggests that we are responsible for healing nature. We who are created in God's image have a responsibility to help guide nature toward that state of wholeness the creator intended for his creation.

Forgetting for the moment that nature has probably never existed in such a perfect state, the basic point is correct that recent damage of nature is of a different order of magnitude than *past* damage and is manifested in ways distinct from nature's *own* processes that create suffering.[68] We have done something to our environment that is unique in nature's history and the natural world is worse off because of it. I assume there are few environmentalists who would dispute this point, though they may argue about whether such changes are appropriately termed "unnatural" or simply "wrong," as Callicott does in denying that global warming is unnatural. If we were to go back a few centuries in history—to preindustrial civilization, say—we would presumably find nature closer to the condition to which ecotheologians wish to see it restored. But this is not all that ecotheologians are saying. McFague, for instance, lapses into eschatological projections of a natural world restored to a state that is not simply preindustrial or even prescientific revolution. Rather, she envisions—and urges Christians to work toward—a natural world in which "there is food for all; where neither people nor animals are destroying one another."[69] This environmental vision calls for the cessation of all destruction of life—even that which is "natural."

This ambivalence about nature's *true* nature is clearly seen in Birch's account of corrupted nature. He rightly condemns many human practices that damage the environment and entail animal abuse: factory farming (which creates excessive soil and water pollution and depletes limited resources, as well as causing animal suffering), zoos, circuses, and other uses of animals as "entertainment," animal experimentation, and destruction of wildlands. On the one hand, he catalogues the anthropogenic causes of environmental destruction and animal suffering and urges us to help put an end to these practices. Yet, on the other hand, he seeks to restore nature and the lives of animals to a state that has never existed in the history of nature. Like McFague, he alludes to a time when all creatures would be "vegetarian . . . they live on grass, and the humans live on nuts and fruit."[70] He looks to the day, foretold in Isaiah, when all of nature "*goes back* to a nonmeat diet."[71] Again, the quest is not merely for nature devoid of humans' worst environmental vices (a difficult enough goal to achieve); rather, Birch seeks to transform nature in light of a biblical vision that has never been, and never could be, matched by biological realities.

Perhaps nowhere is ecotheology's ambivalence about nature—and its hope for a restoration to pre-Fall conditions—so clearly expressed as in the work of Jürgen Moltmann. Recall Moltmann's distinctions between *creatio originalis*, *creatio continua*, and *creatio nova*. The last stage, *creatio nova*, refers to an eschatological restoration of nature to its original state. In this sense, as we have seen, *creatio originalis* already points to "its own perfected completion" in *creatio nova*.[72] Again, like McFague and Birch, Moltmann calls for more than an end to the "progressive destruction of nature by the industrial nations." He wishes to see nature restored to created perfection—a natural world devoid of struggle and suffering.[73]

There are elements of truth in all of these arguments about nature's health and human obligation. But problems arise when environmentalists fail to distinguish "health" (which, as ecologists know, is difficult enough to define) from an original (or future) condition of peace and wholeness that has never existed. Land ethicists are keenly aware of the need to clarify what is meant by land health and to separate appropriate wildlife preservation from the pursuit of a wholesale *redemption* of nature from all suffering and conflict.

Let us first look at the issue of suffering as a natural (as opposed to "fallen") state of affairs. Rolston and Callicott have each written about the misconception rampant in environmental literature that natural suffering is a problem for humans to solve. Both argue that an environmental ethic that takes natural processes seriously (as land ethics does) must recognize the necessary role played by animal pain and death. We encounter different moral requirements for wild animals and humans. Compassion as an ethic may be "morally required" for persons (that is, human beings) but can be "misplaced when applied without discrimination to wild animals."[74] Managing wildlife requires, at times, "a certain callousness toward sentient life," Rolston argues. "One cannot be sentimental about the welfare and sufferings of individual animals in isolation from the inexorable limits of an ecosystem."[75] Compassion for wild animals is not *always* inappropriate, however; the stranded bears in Yellowstone were recipients of a compassionate ethic legitimately applied. Humans, too, are legitimately treated with compassion. Recall Rolston's argument (in contrast to Callicott's) that humans now occupy a postevolutionary position. We are no longer subject to the same selection pressures from nature that wild animals are. Human ethics, including religious ethics, appropriately honors a concern for suffering of other human beings. The feeding of hungry persons advocated in biblical passages (such as Matthew 25:35–40), however, does *not* require the feeding of "wildlife," Rolston argues. "Pointless suffering in culture [i.e., human culture] is a bad thing and ought to be removed where possible"; but the same ethic of compassion is not appropriate in nature because "pain in wild nature is not entirely analogous to pain in an industrial, agricultural, and medically skilled culture. Pain in nature remains in the context of natural selection; it is pain instrumental to survival and to the integrity of the species."[76] In other words, pain in nature must be understood systemically—in terms of ecology and evolution—just as richness and complexity also have to be placed in their broader, environmental, contexts. To intervene where suffering is the result of natural causes, such as a disease, would result in weakening the species as a whole by preserving those members who are not naturally able to resist the disease. Natural pain plays a role in evolution, "even after it becomes no longer in the interests of the pained individuals."[77]

Rolston's understanding of our different obligations to wild animals and to humans reads as a direct rebuttal to McFague's "ecological" ethic of feeding and healing nature. Acknowledging that we have ethical responsibilities toward nonhuman animals should not mean that "humans should treat ground squirrels as persons." Rather, such animals are "to be treated with appropriate respect for their wildness" even when animals suffer as a result."[78] Callicott makes the same point about the inappropriateness of allowing a human dread of pain to dictate an ethic toward nature. He decries the "hyperegalitarianism" of many animal advocates and liberationists.[79] Land ethics does not share the liberationists' "urgent concern" for animal suffering wherever it is manifested.[80] If animals suffer pain, and if (as opponents of speciesism maintain) there are no morally relevant differences between humans and animals, then, some animal advocates claim, we are "morally obliged to consider their suffering as much an evil to be minimized by conscientious moral agents as human suffering."[81] Yet, owing to dread of pain, some animal advocates adopt a position that would be "biologically unrealistic" or even "biologically ruinous" were such ethics actually implemented.[82] "Wild animals," Callicott concludes, "are members of a biotic community" the structure of which is "described by ecology," not human sensibilities. Given these parameters for interpreting and responding to suffering in nature, what does the concept of health mean in ecocentric environmental ethics?

ECOSYSTEM HEALTH AND STABILITY ACCORDING TO LAND ETHICS

It should be noted that ecotheologians are not alone in invoking a concept of environmental health. Advocates of land ethics—Leopold, as well as subsequent proponents—also rely on concepts of land health and aim at a form of preservation or restoration of healthy conditions: a chapter of Rolston's *Conserving Natural Value* is devoted to "Health Values," for instance, and a recently published volume of Leopold's essays, edited by Callicott, bears the title *For the Health of the Land*.

Yet the concept of ecosystem health may not mean health as understood with regard to our own bodies. The idea that an ecosystem can attain, or maintain, a state of health has at times been linked to superorganismal metaphors for nature: just as human bodies are (normally) self-regulating in regard to temperature, metabolism, and other vital functions within the body, so perhaps might an ecosystem exhibit healthy "metabolic," homeostatic processes. Despite skepticism among many ecologists that organismal concepts are accurate or appropriate, the idea of ecosystem health remains useful. However, an ecologist is not a health professional like a doctor who treats pain, heals wounds, and prevents deaths whenever possible, regardless of the cause. In this sense, as Rolston acknowledges, terms such as *health* and *integrity* are "a little strange applied to ecosystems"—and can lead to misconceptions about our ethical obligations.[83]

A thriving culture will always affect nature in some ways, but when this impact becomes extreme, a natural system's health and integrity may be in jeopardy. In order to establish what health means for a given place, it is necessary to establish a baseline of biological integrity "and contrast that with the culturally modified biological integrity."[84] Integrity, in this sense, refers to the ecosystem that was originally there—with the native species that have historically constituted the natural system. Thus integrity has a historic but not a once-upon-a-time, snapshot quality. Integrity is an ongoing process of ecosystems, something that evolves over time. An ecosystem can be said to have health—a "state in which the genetic potentials of an ecosystem's member species are realized"—without having perfect integrity.[85] While such terms have symbolic and ideal dimensions, both *health* and *integrity* can guide environmental policy, setting pollution standards, determining energy cycling, and measuring changes in populations in a given environment.

Leopold himself defined land health as the "capacity for self-renewal in the biota" and linked it,[86] as Rolston does, with ecological integrity, defined as the presence of native species of plant and animal life in the ecosystem: "a high degree of interdependence exists between the capacity for self-renewal and the integrity of the native communities."[87] For Leopold, indications of land sickness include conditions of "abnormal erosion, abnormal intensity of floods, decline of yields in crops and forests, decline of carrying capacity in pastures and ranges, outbreak of some species as pests and the disappearance of others without visible cause, . . . a general and world-wide dominance of plant and animal weeds."[88] Sickness is a relative concept, a matter of intensity and range (or, as Callicott insists, *scale*).

For Leopold, it is true, natural health was inseparable from integrity and stability. Yet as we have seen, some scientists question whether "stability" remains meaningful, in light of developments in chaos theory and the evidence of apparently random population fluctuations, etc. How can we ever know which changes in an ecosystem are "natural" or "normal"? Which changes are bad, given that there is no static, background norm against which to compare "disturbances" to the system? Certainly "equilibrium" ecologists of Leopold's day (the mid-twentieth century) were largely unaware of such debates, and Leopold himself saw little reason to avoid use of such terms as *stability* or *balance*. However, we have also seen that Leopold warns that the "balance of nature" is an expression too simplistic to capture the interactions in nature.

In fact, Callicott argues that, with minor adjustments, Leopold's ethic meets the challenges posed by the "deconstructive ecology" that postdates *A Sand County Almanac* and seems to cast doubt on many ecosystem concepts.[89] Leopold was aware of the dynamic nature of ecosystems and understood that conservation and preservation aim at "a moving target," Callicott argues.[90] A key issue in past and present views of ecosystems is *scale*—a concept that embraces time as well as space. Leopold tended to understand change in ecosystems on an *evolutionary* scale—very slow and gradual—and he may not have realized just how "multiscalar" natural changes can be, Callicott concedes. *Ecological* changes, such as climate or population fluctuations, occur much more rapidly than evolutionary processes such as speciation and extinction. Leopold's focus on evolutionary change may have led him to overestimate the stability and integrity of ecosystems, but he nevertheless incorporated the notion of change, and its normative dimensions, into his land ethic. I would argue, in fact, that there is a sort of multiscalar thinking at work in Leopold's descriptive metaphor of "thinking like a mountain," as when he reflects on the perspective of a mountain overgrazed by deer. He writes that "just as a deer herd lives in mortal fear of its wolves, so does a mountain live in mortal fear of its deer. And perhaps with better cause, for while a buck pulled down by wolves can be replaced in two or three years, a range pulled down by too many deer may fail of replacement in as many decades."[91] Here Leopold delineates, albeit in a poetic fashion, different timescales in nature, moving from the brief lifespan of the individual to the decades or more required for regeneration of foliage pruned by excessive deer herds; the mountain's perspective represents a timescale much longer still than the decades needed for regenerating forests. "Stability," or "health," from the perspective of the mountain, takes account of all these processes, with their different timescales and the relationships between them.

Recall that in Leopold's day ecosystem studies and evolutionary studies were developing in relatively separate spheres; it was some time before a more integrated discipline of evolutionary ecology would coalesce (some would say that a unified ecological theory eludes scientists to this day, in fact).[92] Modern scientists have synthesized ecological change with evolutionary change, arriving at concepts that are coherent with Leopold's basic claim that an action is right if it preserves the integrity, stability, and beauty of ecosystems, wrong when it tends to do otherwise. Land ethics, in other words, merely requires an added measure of "dynamism." Callicott suggests an updated version of Leopold's famous moral maxim: *A thing is right when it tends to disturb the biotic community only at normal spatial and temporal scales. It is wrong when it tends otherwise.* Thus, he argues, we can still evaluate the impact of anthropogenic changes to an ecosystem, even while acknowledging the fundamentally dynamic nature of such systems. Take, for example, the issue of global warming. What, Callicott asks, is "land-ethically wrong with the present episode of anthropogenic global warming?" If humans are part of nature, isn't our "recent habit of recycling sequestered carbon" simply part of our natural way of life?[93] Perhaps so, but we can nevertheless issue a normative evaluation regarding our activities, a judgment rooted in concepts of scale: "We may be causing a big increase of temperature at an unprecedented rate."[94] That, from the standpoint of a dynamized land ethic, is wrong. Human-induced disturbances, judged against a background of natural scales of change, are "far more frequent, widespread, and regularly occurring than are nonanthropogenic perturbations."[95]

Rolston too defends the ongoing relevance of Leopold's insights, despite recent developments in ecology. His account of ecosystem health, like Leopold's, assumes the existence of a *relatively* pristine, baseline for nature against which we can compare human alterations.[96] Herein lies the crux of Rolston's wilderness debate with Callicott, who is skeptical of idealized wilderness and less concerned to separate "things human" from "things natural." Yet even Rolston's more emphatic affirmation of wilderness and ecosystem integrity in no way entails a belief in an original, idyllic state of natural peace and harmony. This brings us back to the criticism raised (though not necessarily endorsed) by Mark Wynn that Rolston's view forecloses eschatological, redemptive improvements to nature. Let us examine Rolston's view of natural health and redemptive environmentalism more closely.

Rolston argues that the real conflict between biology and the Bible lies not in the story of six days of creation but, more specifically, in the idea of a Fall, when a "once-paradisiacal nature becomes recalcitrant as punishment for human sin."[97] He recognizes, as other biologists have (including Darwin) that this story, and the account of evil it assumes, makes no sense in light of biology, which understands struggle and suffering as features of nature long before humans evolved. How might a biologist understand the claim that nature needs to be "saved"? Biologist and ecotheologian Charles Birch argues, for example, that *salvation* is an ecological term, but Rolston is skeptical. He discusses various meanings of the word *redemption*, some biological and some theological. In a nonmoral sense environmental conservation might legitimately be considered a kind of redemption, a way of saving and passing on natural goods. Regenerative and restorative powers of life, he suggests, might also be interpreted as a form of redemption, both at systemic levels and at cellular levels within organisms. (Redemption as regeneration sounds similar to Leopold's concept of health as biotic self-renewal for the land, noted above.) Thus "health" of land may be connected to this understanding of nonreligious, land-ethical redemption: regeneration is a sign of health. Conversely, regeneration might also be linked with *suffering* in natural processes, Rolston notes, as with the birth process. But however it is understood, redemption qua regeneration does not bring an *end* to suffering; rather, suffering is often a "prelude" to regenerative processes. Most significantly, he argues, redemption for nature is not a remedy for inherently sinful, fallen nature. If we wish to apply this term to nature, it "can be redeemed only so far as it has been ruined by human sin"—ruined meaning that humans have created large-scale environmental problems.[98] In this sense the wildest, most untouched places are *less*, rather than more, tainted by sin than cultivated places. And yet it is precisely in wild nature—nature reddest in tooth and claw—that theologians often discern the symptoms of fallenness. Perhaps, as Rolston suggests, "theologians will not be able to figure out what they believe until they have studied biology."[99]

When Rolston describes natural places as "ruined" by humans, he does not mean that wilderness (in the absence of humans) would be edenic perfection, though he does of course maintain that wilderness is a valid concept. His account of nature as "cruciform" and "deiform,"[100] which involves a theological perspective on evolutionary biology, is relevant here. He notes that the presence of struggle in nature creates discomfort for theologians and scientists alike, adding that this term "has largely disappeared from biology texts," being replaced by notions of "adaptedness," and "fittedness."[101] Nevertheless, he argues, the concept of struggle still has an important place in biology. What ought *theology* (as opposed to biology) to say about the struggle for life and the suffering of organisms? Rolston's view of the inextricability of values and disvalues resonates with certain features of process thought. Like Birch and Cobb, he maintains that the unpredictability and randomness of evolution also result in a kind of openness in nature, permitting the ongoing production of projective and emergent values—values implying "possibilities of discovery" inherent in evolutionary processes (134). Like some process thinkers, he holds that evolution is a narrative of life and death, pleasure and pain, constantly intertwined. Prolific evolution "both overleaps death and seems impossible without it" (136).

So, from a theological standpoint, the question one should pose about the existence of such disvalues as struggle and death is "not whether Earth is a well-designed paradise for all its inhabitants, nor whether it was a former paradise from which humans were anciently expelled." Rather, we should ask whether nature shows "significant suffering through to something higher" (142). What does Rolston mean by "something higher?" He is not guaranteeing that a more perfect, compensatory order awaits all sufferers. He regards life as a "passion play" of sorts, one in which "things perish in tragedy" (144). "Redemption" of that tragedy lies not in hopes for an earthly paradise. Rather, the death of organisms brings with it a "passing over" whereby the death itself contributes to, and is part of, other processes of nature (as with genetic load, discussed in the last chapter). Certainly, Rolston discerns a sacrificial element in the ongoing drama ot nature, in the deaths of individuals. But, in his view, even a theological interpretation of redemption does not mean eradication of suffering and other disvalues. "Judeo-Christian faith never teaches that God eschews suffering in the achievement of the divine purposes," Rolston argues. Some Christians may well object that Rolston departs from traditional theological foundations of Christianity in rejecting a vision of nature restored and redeemed. But he concludes that "cruciform creation is, in the end, deiform, godly, just because of this element of struggle, not in spite of it."[102] Rolston accepts a view that is "not panglossian; it is a tragic view of life, but one in which tragedy is the shadow of prolific creativity."[103]

Such judgments about nature, he acknowledges, do not simply emerge from a study of natural history; they are the product of understanding mingled with belief. Interpreting nature in light of theology requires vigilance in order "not to be blinded by one's paradigms, not to accept evil too lightly."[104] But science does play a role in suggesting appropriate theological interpretations of nature. Evolutionary theory indicates that suffering is a constant in nature, yet no species suffers constantly. Rolston raises an issue that was well-known to Darwin: the adaptiveness (or *mal*adaptiveness) of "counterproductive" pain in nature. In many instances, pain is useful, functional; it serves as an alarm system that something is wrong, it motivates organisms to avoid as much as possible situations that involve pain. Extrapolated over generations, or even millennia, pain, and the drive to avoid it, can alter evolutionary history, as when prey species evolve structures or behaviors enhancing their ability to elude predators. Of course, predators too must hone their hunting and killing skills if they are to survive, and therefore pain is never eliminated from the system as a whole. There are also cases in which pain does not appear to be functional, where its excessive or—as Darwin suggests—*depressive* effects make it difficult to locate a clear, biological (much less theological) rationale for it. But instances of counterproductive pain are relatively anomalous in evolutionary theory. Individual organisms may suffer greatly, but on the whole natural selection does not favor constant or excessive pain at the species level: "Any population whose members are constantly in counterproductive pain will be selected against and go extinct or develop some capacities to minimize pain," Rolston argues.[105] In biological terms counterproductive pain tends to be maladaptive and functional pain tends to be favored by selection; in theological terms life as a whole "suffers through to something higher."

In his *Autobiography* Darwin's reflections on the role of pain (and pleasure) in evolution concur with Rolston's. He writes,

> If all the individuals of any species were habitually to suffer to an extreme degree they would neglect to propagate their kind; but we have no reason to believe that this has ever or at least often occurred. . . . Now an animal may be led to pursue that course of action which is most beneficial to the species by suffering, such as pain, hunger, thirst, and fear,—or by pleasure, as in eating and drinking and in the propagation of the species, &c. or by both means combined, as in the search for food. But pain and suffering of any kind, if long continued, causes depression and lessens the power of action; yet is well adapted to make a creature guard itself against any great or sudden evil.[106]

Pleasure both balances pains and is inseparable from them. Most sentient organisms, Darwin suggests, probably experience more happiness than misery in their lives, "although many occasionally suffer much." Such suffering, he concludes, "is quite compatible with the belief in Natural Selection, which is not perfect in action, but tends to render each species as successful as possible in the battle for life with other species, in wonderfully complex and changing circumstances."[107] This, as far as it goes, was Darwin's (secular) theodicy. Darwin had serious doubts that the suffering he described was compatible with theism, since it "revolts our understanding" to think that God's benevolence is "not unbounded."[108] Yet Rolston's account of nature as cruciform illustrates that Darwin's theory does not necessarily undermine theology. "Theologically speaking," Rolston argues, "this position is not inconsistent with a theistic belief about God's providence; rather, it is in many respects remarkably like it."[109]

In a sense, Rolston occupies a middle ground between Darwin and ecotheologians: he does not regard disvalues in nature as an argument against theism, nor does he believe that the eschatological redemption of nature is necessary in order to realign the natural world with the "true," benevolent intentions of God. Like James Gustafson, whose theology we will turn to in a moment, he seriously considers the (apparently radical) possibility that the world—in its present, natural condition,[110] rather than its future perfection—may reflect the will and ordering of God, even while elements of it remain troubling to us. As Robin Attfield observes, Rolston is not proposing that nature is the best of all possible worlds: it might be logically possible to have a system in which there was widespread flourishing of life without the evils of suffering, but such a world might entail "equally widespread supernatural intervention with a frequency that would make nature irregular and fundamentally unpredictable"—even more unpredictable than current ecosystem studies suggest that it is.[111]

Because Rolston does not believe that natural laws themselves are in need of correction, his environmental ethics does not carry biblical connotations of a renewed creation or a perfected community as it does for some ecotheologians. Management decisions, guided by science, "set only strategic not ultimate goals" and attempt to "determine what the spontaneous course of nature was," rather than guide nature into channels favored by human morality.[112] Biblical passages suggesting that nature is fallen ought not to be "taken to suggest that existing wildlands are fallen, nor can they be interpreted in terms of redemptive wildlands management."[113] Redemptive wildlands management is precisely what some ecotheologians propose. But such hopes for nature are misguided when they distort our understanding of what nature is; more important, they obscure the issue of how much and what sort of responsibility humans have toward nature. A "peaceable natural kingdom, where the lion lies down with the lamb, is sometimes used as the symbol of fulfilment in the Promised Land," Rolston notes, but environmentalists must remember that this is "a cultural metaphor and cannot be interpreted in censure of natural history."[114]

To sum up: in land ethics, our ethical obligations to nature do not include eradicating the causes of suffering that are natural. Concepts of health and suffering should be understood within the context of natural processes, including evolution. Although the desire to heal environments whose health has been compromised by human actions points to a worthy imperative, natural process themselves cannot be seen as wrong, evil, or in need of redemption in an eschatological sense. Or, to put the point differently, making natural processes central to environmental ethics suggests a different theological interpretation of nature. Ethics, science, and religion intersect here: the belief that such natural processes are bad in themselves may reflect a theological perspective that supposes nature was once devoid of suffering and death and that evolution itself is somehow symptomatic of human sinfulness and nature's fallenness. This bias, in turn, produces an ethic of redeeming and restoring nature to conditions that are, in effect, unnatural, and such an ethic can only make things worse for the environment we are trying to "save." This ethic misunderstands nature itself and the role that humans ought to play in caring for the natural world. A ecocentric perspective such as land ethics can help to correct these biases. So too can a theocentric perspective.

A THEOCENTRIC CRITIQUE OF ECOLOGICAL THEOLOGY

James Gustafson's critique of Western ethics and theology highlights many of the same problems with ecotheology that land ethics points to. Gustafson's perspective suggests that the anthropocentric core of ethics and theology distorts our understanding of moral action—its possibilities and its limitations. Anthropocentrism constitutes a refusal to accept and respect a natural ordering that is neither of our own making nor completely under our control. To state the point in more positive terms, a theocentric perspective fosters a sense of dependence, awe, and gratitude—or what Leopold terms a sense of wonder or intuitive respect—for powers that sustain human life and life as a whole. Science supports such a perspective, Gustafson argues, because it reinforces the idea that humans are not the center of these powers and processes. This too, I take it, is Leopold's point when he argues that knowledge of nature gives us "a sense of wonder" about the biotic enterprise of which we are a part but not the "sole object."[115] In the next chapter we will discuss in greater detail the theocentric and land ethics interpretation of "participating" in natural processes rather than trying to control them. For now, I want to focus on Gustafson's critique of anthropocentric biases in theology and ethics and see how his criticisms supplement and reinforce those articulated by land ethicists such as Rolston and Callicott.

Thus far the criticisms of ecotheology that I have developed have taken their bearings primarily from secular (philosophical, scientific) perspectives, such as Darwinism and land ethics (though, clearly, Rolston discerns theological meaning in the biological ordering). Yet the issues ecotheologians raise about the natural world—issues of justice, suffering, redemption—are, of course, not merely empirical or scientific matters. These are issues of theology and theodicy. Thus ecotheologians might well object that one cannot criticize their arguments from a purely secular/scientific perspective—it is not the role of science to test the legitimacy of theological propositions. The claims of theologians are of a different nature than those of scientists. It is worth recalling, however, that each of these ecotheologians maintains that the accounts of nature they present are informed by and consistent with natural science. My intention is not so much to test their theological claims with exclusive reference to scientific data or secular perspectives but rather to illustrate how a scientifically inaccurate understanding of nature perpetuates ethical and theological imperatives that are inappropriate. The belief that humans are obligated to reduce suffering and restore peace in nature is an example of the way in which an inaccurate or incomplete picture of nature leads to questionable environmental action.[116] In any event, there are problems with ecotheology that go beyond its scientific inadequacies. A *theological* critique of ecotheology is also indicated and Gustafson's work serves as a touchstone in this endeavor.

In *A Sense of the Divine* Gustafson argues that "to live and act in a theocentric perspective is to know and feel the relativities of all other centers of value proposed by my fellow interpreters of the natural environment."[117] To construe nature and our place in it from a theocentric perspective is not to claim that we can construe it from *God's* perspective, nor does Gustafson believe that we can leave behind our own perspective as human beings—on the contrary, our position as finite beings in a particular place and time reinforces the need to construe ethics theocentrically.

Gustafson argues that the majority of current theology and ethics holds that "material considerations for morality are to be derived from purely human points of reference." Questions of what is good or valuable take the form of "What is good for [humans]? Or "What is of value to human beings?"[118] In Christian theology the anthropocentric concentration is seen in a persistent preoccupation with the well-being and salvation of individual humans or human communities. "Religion," Gustafson contends, "is increasingly advanced as instrumental to subjective temporal human ends," including the desire for "getting on well in the business of living, for resolving those dilemmas that tear individuals and communities apart, and for sustaining moral causes."[119] A focus on issues of personal repression and oppression, he laments, has come to replace "traditional" concerns with the problem of sin and guilt. Religion is promoted as a means to personal growth, a therapeutic response to anxieties. "One should become liberated from whatever represses or oppresses," according to much of contemporary teaching.[120] This focus on human salvation values God in terms of utility and therefore God is denied *as God* and becomes merely an instrument for fulfilling the hopes and alleviating pains and fears of human beings.

The focus on human beings as the primary point of reference in ethics and theology also flies in the face of centuries of well-established *scientific* knowledge. "Theologians and ethicians have shown remarkable myopia," Gustafson observes, "in not taking into account the inferences that can reasonably be drawn from some of the most secure knowledge we have of the creation of the universe, the evolution of the species, and the likely end of our planet as we now know it."[121] He proposes that ethics and theology consider a different set of questions that attempt to discern what is good for a larger whole: the human species, other species, or nature broadly construed.

But discerning what is good for this larger whole is itself fraught with ambiguity. It is inevitable that conflicts will emerge among various communities or between communities and individuals; our moral actions are constrained by past actions as well as by our nature as finite beings. We cannot perceive or comprehend the larger whole whose good we seek to promote, and for this reason our ethical choices will necessarily be difficult and complex. "The multidimensionality of value, or values," Gustafson argues, "casts us into ambiguities of choices that are unavoidable."[122] Ethics involves us in the task of trying to discern the relationship of parts to wholes and making decisions based on this (imperfect) discernment. While ethical decision making attempts to reduce ambiguity and to negotiate conflicts, it can rarely, if ever, completely resolve them. To assume that conflicts can be harmonized and that their resolution will necessarily favor human interests and concerns implies a refusal to consent and conform to an existing order that sustains a whole larger than ourselves. Anthropocentrism denies God as the "power and ordering of life in nature and history which sustains and limits human activity, which 'demands' recognition of principles and boundaries of activities for the sake of man and of the whole of life."[123] Realizing our situation as beings at once dependent upon, and limited by, the given ordering is not an argument for moral passivity or withdrawal from decision making, Gustafson argues. As moral agents, we are subject to the dual conditions of "finitude and responsibility."[124] In short, we can neither resolve all conflicts nor simply refrain from engaging them.[125]

As I suggested above, a theocentric construal does not mean viewing the world from *God's* point of view but rather acknowledging "that the source and power and order of all nature is not always beneficent in its outcomes for the diversity of life and for the well-being of humans as part of that."[126] We should relate to all things in a way that is appropriate to their relations to God. God is the Other, the ultimate reality, in relation to whom "we are to relate ourselves and all things."[127] The sense of dependence people experience in connection with this ultimate reality may or may not be a distinctly religious affectivity, Gustafson concedes. Theology is merely one way of construing the world, of interpreting the relation of things in the world to the conditions that make them possible. Many people, religious and nonreligious alike, experience a dispositional response of gratitude for nature as a power that "bring us and all things into being, powers we did not create, and some of which we cannot yet control."[128] But for a religious person these powers "manifest the power of the ultimate reality," understood as God.[129] The conditions of finitude and dependence within which we make moral decisions exist whether we respond to the given ordering of the world with a religious or secular affectivity. "This dependence—a matter of fact, no matter how it is interpreted—evokes a sense of the sublime, or for some of us a sense of the divine."[130] Ethics requires some process of discernment of an objective ordering of events. For his own part, Gustafson responds to this ordering with religious feelings of dependence, awe, and gratitude.

This summary of Gustafson's theocentric perspective certainly does not encompass all the important elements of his argument (and some, such as his view of interdependence and conflict will emerge again in the next chapter). But it is, I hope, inclusive enough to allow some comparisons to be drawn between theocentrism and the perspective of some ecological theologians. In drawing these comparisons we will focus primarily on the issue of anthropocentrism in theology and the problem of "instrumentalization" of religion and God to which Gustafson alludes.

One might expect that ecological theology, by its very nature, would take a more expanded view of life and assign to humans a less central role, in light of ecological and evolutionary considerations. To some extent this is in fact the case, but their project remains incomplete. Just as Rolston laments the lack of a "paradigm shift" toward a less anthropocentric perspective in environmental ethics, Gustafson's critique points to a similar failure to turn theology away from an anthropocentric concentration. While the starting point of any theology will necessarily entail human perspectives, the subject matter of theology need not be limited to human beings and their experiences. Values can be generated by humans without necessarily being centered on them. "Man is the only measurer of things," Rolston argues, "but man does not have to make himself the only measure he uses."[131] Gustafson observes this distinction as well when he acknowledges that a certain kind of anthropocentrism can never be completely avoided in ethics. The human species, he argues, "is always the *measurer* of all things" but this does not necessarily imply that we are the "measure."[132] The tendency to view humans as the measure, not just the measurer, Gustafson notes, is exemplified by religious and ethical "preoccupation with subjective problems of individuals."[133]

Ecotheologians understand this distinction between anthropocentric and anthropogenic orientations as well, but the distinction seems to get lost immediately after it is drawn. McFague for instance acknowledges that theological interpretation always emanates from a particular point of view—the point of view of humans in a given place and time. Nevertheless, her intention is to suggest metaphors and practices that are nonanthropocentric, a "theology that is creation-centered, in contrast to the almost total concern with redemption in some Christian theologies."[134] The problem, then, is not so much the source of theology and ethics from human perspectives but the intended subject matter. I want to argue that while much of ecological theology succeeds in broadening its scope of life to include nonhuman forms, it does not always succeed in discerning the value of nonhuman life in nonanthropocentric ways. With this distinction in mind, let us see what Gustafson perceives to be the markers of anthropocentrism in contemporary theology.

Gustafson identifies several strands of contemporary religious thought that illustrate a trend toward utility and anthropocentrism. His critique includes some forms of liberation theology, eschatological theologies (such as Moltmann's), and (to a much lesser extent) process theology. He discusses each of these branches of theology in terms of their common or more "traditional" forms—that is, he does not attempt an assessment of these theologies as they have been developed along ecological lines, though all of them have recently taken such a turn.

As with his criticisms elsewhere, he cites the narrow view assumed by most theologians that humans and their concerns are the primary focus of theological inquiry. We will begin with process theology and then look at some of Gustafson's remarks about liberation and eschatological theology. Clearly, there are certain areas of overlap in these theologies (the goal of liberation, concern with suffering) and therefore some of the same criticisms apply to each. In some instances, particularly in the discussions of process theology that follow, my criticisms are *based* on Gustafson's critique of theology, even though he does not arrive at the same judgments as mine in each case. This is particularly so with liberation theology, which we turn to first.

PROCESS THEOLOGY

Gustafson finds process theology considerably less anthropocentric, and less problematic in general, than many other kinds of contemporary theology. He notes that, among process theologians, John Cobb has contributed the most to establishing a clearer link between the often abstruse language of process thinkers and a practical application of ethics (Whitehead himself was more concerned with aesthetic than ethical language.) Gustafson acknowledges that there are certain affinities between process views of God and nature as "relational" and his own theocentric account. Both perspectives imply a discernment of relative values in specific contexts. Moral action in process theology proceeds from an investigation into the context, the relationships among values, in a given situation. "Prior to any obligation to actualize or realize values is their emergence; prior to any action to realize them is the appreciation of them."[135] This process resembles Gustafson's approach to ethics, which posits a "multidimensionality of values and relational value theory."[136]

Gustafson argues that process theology tends to be less anthropocentric than many other approaches. It offers fewer assurances about ultimate outcomes (i.e., eschatological outcomes) that are beneficial to human beings than a theology such as Moltmann's. Furthermore, process thought understands God as "related to the natural world," not merely to the world of human intentions and actions.[137] However, one of the primary problems with process theology in Gustafson's view (aside from its tendency toward abstractness and generalization) is that it relies upon an almost exclusively rational, rather than affective, judgment of the dependence and interdependence of all life. The relationality of all things—and thus knowledge of God as well—is rooted in rational human experience. In other words, Gustafson commends process thought for what he sees as its essentially theocentric approach, but he is not comfortable with its epistemology.

My own criticisms are informed by Gustafson's remarks about the role of rational experience in process thought. However, based on my reading of process theology, and in particular the work of Birch and Cobb, it is difficult to see how its account of the appreciation and realization of values can be considered essentially theocentric. As we have seen, the values to be appreciated and realized by means of Birch and Cobb's richness of experience criterion are highly anthropocentric. Furthermore, because process theology draws on the paradigm of *liberation* of life, it *does not* fully succeed in "turning religion from preoccupations with benefits."[138] Rather it simply extends those benefits to both humans and (certain) nonhuman animals—namely, those animals most like humans, based on the richness criterion.

Gustafson notes that process thought makes human rational experience central, but his complaint seems to be more with its concentration on *rational* experience than on *human* experience. Gustafson's uneasiness about the epistemological basis of process thought derives from its insufficient attention to an affective, rather than purely rational, response. As I have argued previously, the problem I see with process thought (at least as a basis for environmental ethics) is that value in nonhuman life is assessed in terms of categories of experience (largely rational, as Gustafson notes) that take human experience as their reference point.

The problem with process thought is not just that it posits humans as the knowers or measurers of all things, including God, but that it also makes humans the measure. Birch and Cobb's hierarchy of value in animal life makes it clear that human value is understood as the baseline for all other value in life. As Northcott argues, the emphasis on locating similarities between humans and nonhuman life elides the "genuine difference or otherness between humans and nonhumans, plants, cells and atoms" encouraging a "homogenising view of the natural world which may simply encourage its further homogenisation."[139] The homogenization is a reflection of process thought's preoccupation with human experience and its value as normative for all life.

For such reasons I disagree with Gustafson that process thought (at least Birch and Cobb's formulation of it) is largely theocentric in orientation. In characterizing the anthropocentrism that persists in theology, Gustafson remarks that in much of this literature the

> dignity of the [human] species is protected by those precepts which require that we favor man over animals in all instances in which human life is threatened . . . those precepts which make it clear that the natural world can and ought to be put in the service of human aspirations for health and reasonable material comfort, for development of culture, and for fulfillment of individual well-being.[140]

Such precepts are embedded in Birch and Cobb's evaluation of animal life vis-à-vis human. As we have seen, they discuss specific conflicts between human well-being and the proliferation of nonhuman species (as in the case of conflict over land between Rwandans who want to grow crops to feed their expanding population and elephants for whom the land is habitat). Birch and Cobb concede that "Ruanda would be morally obligated to abandon its protection of elephants."[141] "Such choices are never easy," they add. Yet, the richness criterion makes them easier than they ought to be.

I am also somewhat less certain than Gustafson that process theology's construal of God's relation to the world is immune to criticisms of it as essentially utilitarian. As noted above, Gustafson is critical of theologies that understand God as a utility device for the enrichment and fulfillment of human goals and desires. A theology oriented toward guaranteeing certain benefits for humans not only instrumentalizes God, he argues, but implies that we do not need to consent to an existing ordering. In process thought assurances about liberation from conditions of life (including suffering entailed in natural processes) are a central and problematic feature; moreover, God is clearly understood as a means to personal fulfillment. Liberation of humans as a whole, and of individual oppressed persons, are primary goals in Birch and Cobb's theology. A few examples will illustrate this point.

Birch and Cobb repeatedly denounce the dominant paradigm of oppression that has tended to "treat living organisms as objects to be manipulated rather than subjects that experience" (1). Believing that we "*can* free ourselves from the tyranny" of this paradigm, they "call for the liberation of life" (2). By stressing that all things are constitutive of their relations with other living things, the ecological model deconstructs objectification and dissolves dualisms (91). Liberation in this sense involves a shift in our thinking about life. Recall that Birch and Cobb also have a second, more emphatic sense of liberation in mind, however—that which proceeds from maximizing richness of experience. In dealing with nonhuman animals, our obligations are primarily to reduce suffering or avoid increasing it, but with human beings, the goal is a more positive one, involving increasing their richness.[142]

What does it mean to enrich the lives of other humans? Birch and Cobb suggest adopting an attitude of "trust in Life" where Life "as the central religious symbol is God" and trust implies "allow[ing] the challenging and threatening elements in our world to share in constituting our experience . . . believ[ing] that they can enter into a creative interchange with what our past experience brings into the situation."[143] Clearly, Birch and Cobb argue, trust in Life cannot simply mean "that one trusts Life to support one's present projects and guarantee one's success," which would lead to personal stagnation (since humans need transcending experiences; ibid.). But such trust does "enrich" us and make us "more alive" (184). While trust in Life is not a formula for unmitigated happiness and success, it does translate into something like human "growth." In this sense, trust in Life clearly involves an interpretation of God as a source of benefits to humans and a general preoccupation with the experiences of individual human subjects.

The examples of trusting Life that Birch and Cobb offer underscore this point: one involves an "alcoholic who joins Alcoholics Anonymous and through the interaction with other members sobers up" (183). Birch and Cobb warn that such a person should not perceive his or her sobriety as a "final" good but should remain open to the future "work of Life." Another example describes an artist who achieves "artistic freedom" after a long process of "arduous discipline" (185). Trust in Life can also help those in the West to overcome the "estrangement from our bodies" as objects. They hope that trust may yet lessen our anxious desire to control other living things as well. To sum up: process theology, according to Birch and Cobb, is in fact largely preoccupied with benefits for humans, including the removal of their anxieties and liberation from a personal sense of repression and oppression. Birch and Cobb extend some of the benefits of liberation to animals as well—provided they demonstrate certain similarities to us. The overall goal is personal liberation of human subjects and the liberation of animals who share our capacities of subjecthood. Gustafson does not take issue with (or is not familiar with) these particular features of process thought as Birch and Cobb present it. Nevertheless, these criticism follow from his analysis of anthropocentric and utilitarian forms of theology and ethics. Some of these problems are also apparent in liberation theology, which is the topic we turn to next.

LIBERATION THEOLOGY

Gustafson does not dismiss the goals of liberation theology outright. He acknowledges that on the whole it is a "theologically serious movement" that seeks to bring about important social change, but his concern is that it often puts God in the service of the interests of one particular community.[144] He wonders whether liberation theology's concept of a "creative, liberating deity is sufficient for an adequate understanding of the ordering power that ultimately limits and directs history and nature"; God appears to be limited in his actions "too exclusively to the realm of history and not sufficiently to the ordering of nature."[145] The central claim here is that much of theology is oriented toward human social goals and historical developments, while it is insufficiently concerned with the natural ordering as a context in which these goals must also be defined and evaluated. Nature provides us with the "materials and occasions for human creativity," but it also threatens us and provokes awe and respect.

Liberation theology tends to refashion God and the natural ordering in the shape of human expectations, molding theology into a direct response to a human set of concerns—about social oppression, for instance, or physical suffering, or personal tragedy. It assumes that all such tragedy can be overcome, "or if cannot be overcome, it has a purpose which is *realizable in benefits to individual persons and to the species.*"[146] Such use of theology borders on idolatry in Gustafson's view, implying a repudiation of God as God.

One example Gustafson cites of utilitarian use of, and language for, God is the "blatantly trivial uses of piety" to promote social and moral ends seen in statements such as "'God is an unwed mother on the West Side of Chicago.'"[147] Such characterizations of God, he concedes, are intended to "shock" Christians into recognizing their moral responsibilities and failures regarding others. He recognizes that language for God is often metaphorical or analogical; nevertheless, he considers certain uses of religious language wrong and, at a certain level, even blasphemous.

McFague's liberation theology-oriented form of "Christian nature spirituality" would appear to be a case in point of employing metaphor in order to shock Christians out of their typical modes of thought and expression. She holds that just as Christians see the face of Jesus in "the needy brother and sister" so too should they recognize "that face in a clear-cut forest, an inner-city landfill, or a polluted river."[148] McFague does not go so far as to say that Jesus *is* a landfill, but her point is to call attention to the similarities between Jesus and suffering nature: nature too "is being crucified" (174). This usage is distinct, in my view, from Rolston's account of nature as cruciform, for various reasons already explained but primarily because Rolston does not see suffering as a form of "oppression" ultimately overcome or (as Gustafson puts it) as having a purpose that can be realized in benefits to individuals and communities. In McFague's account, once the connection is made between nature and Jesus Christians understand that their moral responsibilities have a "particular slant"—toward solidarity with the oppressed. Nature "is joined in its oppression with Christ" (ibid.).

The crucial difference between McFague's uses of theology here and those that Gustafson's theology explicitly rejects is that she has not put theology in the service of human needs *alone*. Clearly, she intends to extend moral concern beyond the restricted range of human desires and expectations as well as to expand our language about God. Yet, despite the fact that McFague sets out to consider the question "What is good?" from the standpoint of both humans and nature, she is not able to break out of a human frame of reference; hence her concern for the natural world as an "oppressed" class (itself a category derived from human social struggles) regards nature as the "new *poor*," lamenting the "increasing poverty of nature" (170–71).

As an ecofeminist McFague condemns the negative association of nature and women that has led to the objectification and control of both by (usually, but not exclusively) powerful white males. "The depiction of female nudity is a key marker in the development of the arrogant Western eye toward *nature*," McFague argues, "because of the old, deep, and pervasive relationship between women and the natural world" (81). Nature, "like women," has been regarded as a mere "object of the arrogant eye." Each of us in Western society is guilty of this objectifying gaze, which "organizes everything in reference to oneself" and cannot imagine true otherness (33).

Like Gustafson, McFague is aware that some degree of anthropocentrism is inherent in talk about God and religion. She grants that all knowledge, all seeing, is inherently "perspectival"—we always start with *some* frame of reference, whether ethnic, cultural, or economic. Yet, the "alternative" perspective on nature that she offers does little to correct the problems of the old paradigm. Note how McFague's ecofeminist critique simply shifts the focus from one human reference point (nature = women) to another (nature = the oppressed/impoverished), never leaving behind the narrow perspective that, by her own account, obliterated nature's distinctiveness in the first place. Indeed, the basic association of nature with women is ultimately reinforced; she replaces an oppression praxis with a liberating one but does not succeed in granting nature a significance or value separate from human experiences and concerns. (This is evident in McFague's more recent work, *Life Abundant*, in which she claims that ecological theology *is* liberation theology.) "A symbol of [nature's] deterioration," she writes in *Super, Natural Christians*, "is a poor third-world woman of color, for she is a barometer of the health of both humanity and nature . . . In her increasing poverty, we see also the increasing poverty of nature" (171). McFague conceives of nature in evaluative categories that are not only human-centered but in fact speak to a very specific set of human economic and social concerns. In this way she fails to appreciate the very otherness of nature—its "mystery, complexity, and difference"—that her alternative perspective is intended to preserve and respect. "We can scarcely imagine what it would mean for nature to be considered Other in the sense of being independent of and indifferent to human interests and desires," she warns (34–35). Such imagination is indeed hard to come by, and McFague's account of nature will not bring us any closer to it.

A similar theological objection can be lodged against Rosemary Ruether's ecofeminism. We have seen that Northcott characterizes Ruether's ecotheology as "humanocentric" (rather than "theocentric" or "ecocentric"). The humanocentric orientation of her work is apparent, he argues, in the way that Ruether "constructs her ecotheology on the basis not of a new account of nature, nor a new account of God, but of her critique of patriarchy and the ecofeminist account of the normative significance of the experience of women for human relations with the non-human world."[149] Ruether starts from a human experience—in fact, the particular experience of women as oppressed—and proceeds to interpret nature in light of it. Wishing to see relationships of domination and oppression eradicated, Ruether interprets nature as a kind of ideal community in which mutuality and cooperation reign. Human forms of competitiveness are qualitatively different from nature's, causing damage to ourselves as well as all other life-forms. Her description of nature expresses the ecofeminist longing for a world without oppressive structures. In Gustafson's terms, it reinterprets the existing, created order in light of human experiences of social oppression, sexism, and racism. These criticisms point to the same conclusion that our earlier, scientific evaluation of Ruether's claims also suggested: her ecological ethic is derived not from nature, as she claims, but from human communities. Again, we see that the *scientific* flaws in environmental ethics are intertwined with a theological orientation that remains anthropocentric.

Moltmann's theology presents us with a final case in point of the persistence of anthropocentrism in theology and the tendency to put God, and the natural ordering that manifests God, in the service of human hopes and desires.

MOLTMANN'S ESCHATOLOGICAL THEOLOGY

Gustafson takes issue with Moltmann's understanding of a renewed creation (*creatio nova*) and the centrality of the concept of resurrection for his interpretation of nature. Much of eschatological theology, he argues, "continues to assure human beings that the Deity serves to fulfill particular human desires." In Moltmann's theology this orientation is reflected in the central symbol of Christ's resurrection which is assumed to imply "that nature is open to a new creation."[150] Moltmann's God, in other words, is expected to reorder creation in ways that better conform to human hopes. We have previously seen that *creatio nova* represents a different sort of created world than the one we now inhabit. Again, these theological claims, in Gustafson's view, constitute a "denial of the need for humans to *consent* to 'being'"—that is, to consent to the "power and ordering of life in nature" (1:84). The denial of the given order in Moltmann's argument in favor of a new creation (established by the spirit of God who dwells in creation) expresses his preference for a world devoid of evolutionary forces that produce struggle and strife. In *God in Creation* the spirit of God is assumed to banish all negativities of the natural world. We see once again that an anthropocentric perspective correlates with an inadequate and incomplete understanding of natural processes such as evolution.

It should be noted that Gustafson's criticisms focus on some of Moltmann's earlier work. Moltmann argues in his *Theology of Hope* (1967) that we should approach nature, and indeed all reality, with an attitude of hope—hope stemming from "the assurance of the possibility and actuality of a new creation" (1:44). Gustafson wonders why a new creation would not produce more "dread" than hope, given that we have come to rely on nature's processes and patterns as they are. Moltmann's reason for an attitude of hope rather than fear, however, lies in a belief that "the content of the resurrection assures us that it is beneficial to the human species" (ibid.). A theocentric perspective suggests that such belief in a radical new creation not only contradicts "centuries of development in the natural sciences" but reveals a persistent conviction in Moltmann's theology that "the whole of this universe was created solely for the sake of man" (1:44–45).

In Moltmann's more recent works, such as *God in Creation*, his understanding of the "new creation" is explicitly broadened so as to extend eschatological hope and renewal to *all living things*. In this sense, he cannot be accused of restricting his theology to an entirely human realm of existence. As we have seen, he develops an account of our current ecological crisis and of the evolutionary processes in nature that are part of *creatio continua* ("the theory of evolution," he notes "has its place where theology talks about continuous creation").[151] Human beings, he urges, "are not the meaning and purpose of evolution."[152] Rather, we must understand human history in the context of "the great, comprehensive ecosystem 'earth'" and, in order to do so, it is necessary to "refrain from drawing a dividing line between the theological doctrine of creation and the sciences and their scientific theories."[153] In other words, Moltmann appears to have modified his theology along the very lines of Gustafson's critique of anthropocentrism: "We have to overcome the old anthropocentric world picture by a new theocentric interpretation of the world of nature and human beings, and by an eschatological understanding of the history of this natural and human world," Moltmann contends.[154]

Despite these adjustments, Moltmann has not significantly altered the anthropocentric orientation of this theology.[155] While he speaks more broadly of creation in general, his theology retains the basic goal of eliminating those conditions in life (both human and animal life) that generate suffering and conflict. The spirit in creation continually works to adapt each organism to its environment until the time of renewed creation when the yearning of all life for liberation is fulfilled. Like Birch, Cobb, McFague, and Ruether, he expands the scope of his theology to include nonhuman life but ultimately fails to interpret life as a whole from a standpoint that is distinct from human expectation and desires. Interpreting God and nature theocentrically does not mean that one has to be utterly indifferent to issues of suffering and well-being in human or animal lives but, as Gustafson points out, "it does set these concerns within a different and wider theological context."[156] Broadening theological contexts beyond their traditional anthropocentric bearings may be disorienting. It suggests, among other things, that in "forces of nature, conflict and destruction mar the harmony that is romantically appealing."[157] However unappealing the perspective may be at times, a theocentric construal does not force God and nature into roles that better suit our own preferences for harmony and justice. "If there is a sense of divinity," Gustafson argues, "it has to include not only dependence upon nature for beauty and sustenance, but also forces beyond human control which destroy each other and us."[158] Moltmann's account of the stages of creation assumes that God necessarily shares his particular hopes for the casting out of all forces that create struggle and strife in human and nonhuman life.

In conclusion, Gustafson's criticisms—and criticisms informed by his work—highlight the narrow, human-centered perspective that dominates ecotheology, even while these authors strive to include nonhuman forms of life. Human categories of experience and value are transferred to nature, and the basic ethic toward human and nonhuman life is assumed to be the same. This is not to say that theology and ethics should not be concerned with human rights, suffering, oppression, sexism, and so on. As Gustafson acknowledges, there are "ample grounds in the biblical materials for being concerned about the poor and the oppressed."[159] But it is to say such values and objectives cannot simply be extended to nonhuman life. The questions that a truly theocentric ethic would put to ecotheology are "Is the Deity solely on one side of the issue? Can God's power be understood to be more sovereign, thus ruling nature as well as social experience, and thus qualifying the certainty of his identification with one cause or one course of action?"[160] Ecotheologians seem to embrace God's action in the realm of history alone; the natural order is assumed to conform (or be clearly subordinate) to the historical sphere, and the human pursuit of what is perceived to be good is not seen as qualified by the natural order in any significant way.

Theocentric ethics and land ethics recognize the conflict of values and the inevitability of tragedy that are part of nature and that follow from taking natural processes seriously. Gustafson argues that the natural ordering "brings us into life and sustains life," but it also "creates suffering and pain and death."[161] Nature as *a whole* is to be respected, even when its ordering does not seem to support what we see as desirable outcomes. Leopold, Callicott, and Rolston similarly describe the wonder we experience at the biotic enterprise as a whole, even when its processes entail struggle and death for individual beings.

Conflicts are inevitable in ethical decision making as well as nature itself, and we cannot alter the natural world and its inhabitants to suit our own moral preferences. Recognition of this fact suggests a more modest role for humans in our dealings with nature—we are, in the language of land ethics and theocentrism, *participants* in natural processes and patterns that are not of our making. The concept of humans as participants will be more fully explored in the next chapter. We will consider what a more limited ethical role for humans means in terms of traditional Christian ethics, and particularly a dispositional ethic of love and gratitude toward nature.

CHAPTER 6

A Comprehensive Naturalized Ethic

A man with the milk of human kindness in him can scarcely abstain from doing a good-natured thing, and one cannot be good-natured all round. Nature herself occasionally quarters an inconvenient parasite on an animal towards whom she has otherwise no ill will. What then? We admire her care for the parasite.

—George Elliot, *The Mill on the Floss*

In a recent work entitled *God After Darwin* John Haught observes, "To a great extent, theologians still think and write almost as though Darwin had never lived."[1] I have argued throughout this work that much of contemporary ecological theology is a case in point. The failure to take nature seriously is particularly problematic in environmental ethics where knowledge of natural processes is essential. But what does it mean to take nature seriously? Can we, and should we, follow nature in some simple and straightforward manner? Should ethics follow directly from evolutionary and ecological considerations? I have claimed that it should not, while maintaining that scientific knowledge must play a central role in the process of discerning our ethical obligations toward nature. In the pages that follow I argue in favor of a "comprehensive naturalized ethic," borrowing a phrase from Holmes Rolston. In developing this argument I will use the term *nature* in different senses: our own human "nature" as well as nature in the broader sense, as the sum total of natural processes and other life-forms in the world. There is, of course, a relationship between "human nature" and nonhuman nature; I will also explain what I think this relationship is.

Keeping these interpretations of nature (in various senses) in mind, I will then turn to the task of formulating a better environmental ethic—one that takes natural inferences as well as theological orientations into account and locates common ground between the two. The alternative developed here revolves around arguments from land ethics and theocentric ethics that, as I have already suggested, share a similar perspective on the role of humans in nature and the qualified moral obligations that our position entails. In this chapter we return to an issue raised by some ecotheologians: whether it is appropriate to regard nature and nonhuman animals with an attitude of love. It is possible to extend an ethic of love toward nature and animals, I will argue, provided that the ethic is qualified in important ways that I outline, in an admittedly preliminary way, toward the end of this chapter. In the course of defining the parameters of an appropriate ethic of love toward nature, we will look at two arguments that attempt to explain human affective responses to nature from a biological perspective. While both of these arguments offer some promising possibilities for developing an environmental ethic based in love, they also suggest that we encounter obstacles to extending such an ethic widely, to all aspects of the natural world.

With these caveats in mind, I wish to argue for a more limited and less interventionist kind of love toward nature—one that recognizes the reality of evolutionary kinship with animals as a partial basis for love, while acknowledging that we need to be *discriminating* in how we extend such an ethic. An ethic of love defended on the grounds of evolutionary continuity and sameness between humans and animals is not sufficiently discriminating. Loving responses must be carefully defined with reference to the kinds of beings we are dealing with. In setting the parameters for an appropriate, minimally interventionist love ethic toward nature, I will draw on some of Bishop Joseph Butler's reflections on human moral responses and the proper objects of love, linking these to the insights of philosopher Mary Midgley. The perspective of land ethics as articulated by Leopold, Rolston, and Callicott also suggests a minimally interventionist ethic toward nature. A less interventionist, naturalized ethic is likewise indicated, I believe, by Gustafson's theocentric approach.

To be sure, ecotheology already constitutes a naturalized ethic of sorts. Ecotheologians would quickly point out that their ethics toward animals and the environment have taken scientific data and inferences from nature seriously. We have seen that the "scientific" concept of nature's interdependence plays a central role in environmental ethics and, for that reason, we will begin by looking more closely at this concept and its use in ecotheology. Descriptive or factual interdependence glosses over the conflicts that naturally exist in biotic systems and moves directly to an ethic of interdependence as the key principle to be derived from the study of nature. This in itself is problematic. But it is also important to note that interdependence is used in at least *two senses* in ecological theology and it will be necessary to distinguish these senses before determining the extent to which they adequately reflect natural processes.

INTERDEPENDENCE IN ECOTHEOLOGY

As we have seen, among environmentalists, and particularly ecotheologians, the concepts of interdependence, interconnectedness, and interrelationship form the backbone of ethics. These three terms are often used interchangeably to describe the links that exist—biological, moral, even causal—between humans and all other life-forms.[2] The idea of interdependence is offered as a corrective, a better alternative to relationships of domination, oppression, and modes of thought that are objectifying and dualistic. Interdependence promotes solidarity, points to our similarities—our kinship—with all other life-forms, and also situates us within a larger network of beings. Thus "interdependence" may be used in at least two ways.

The first is the sort of interdependence (perhaps best described as inter*connected*ness) expressed in ecosystemic relationships, wherein the activities and fate of one member of the system have consequences for all others. Interdependence in this sense is what ecotheologians have in mind when they speak of our own nature as constituted by our relationships with other living things; our species, and indeed every species, has been modified in the context of all other organisms with whom we have come into contact. At any given time this process is happening in an ecosystem as organisms are acting and being acted upon. Interdependence of this sort is also commonly expressed in terms of food chains in nature or the web of life. We have seen that Leopold, for instance, refers to chains of dependency constructed of trophic relationships between predators and prey.[3] Animals engaged in this sort of interrelationship are not necessarily close relatives biologically and may in fact be quite distantly related. As Darwin describes it, "Dependency of one organic being on another, as of a parasite on its prey, lies generally between beings remote in the scale of life."[4] Kinship, in other words, is not foremost in this sort of relationship, though in the broadest sense all organisms can be said to be related.

A second kind of interdependence (more precisely, inter*related*ness) consists in the genealogical, evolutionary continuity that exists between all living things.[5] We are kin with all other life-forms, given our common evolutionary heritage. Whereas ecosystem interdependence is not necessarily a function of biological kinship—interactions occurring within the system can take place between organisms that are very closely or very distantly related biologically—this type of interdependence expresses literal kinship with animals. Interdependence as biological continuum stresses our genealogical relationship—shared genetic material—with other organisms. Both kinds of interdependence are assumed by ecotheologians to express scientific (at times, specifically Darwinian) "facts" about the world, although the first is perhaps more ecological in its emphasis, the latter more evolutionary. In discussing these two concepts of interdependence, I will make sharper distinctions between them than ecotheologians often do, while still acknowledging that they are in some sense inseparable in light of contemporary evolutionary ecology.

Ecotheologians tend to treat the idea of interdependence as though its positive ethical meaning were self-evident. Or, put differently, they tend not to separate the "fact" of interdependence from the "ethic" of interdependence. Interdependence is offered as both descriptive of nature and prescriptive for our treatment of it. It is assumed to be an objectively good or desirable thing, in the same sense that "community" is assumed to be good, without much further inquiry into what natural interdependence actually entails. This move from the fact of interdependence to an ethic based in this fact (an ethic of love or community or mutuality) is made too quickly, as when McFague refers to the "*evolutionary, ecological, relational, community* model of nature" without pausing to tease out the distinct meanings of each of these key terms.[6] "Ecological ethics," Ruether concurs, "arise from the description of how nature works."[7] Let us turn to a more in-depth analysis of natural interdependence, both in its (ecological) "systemic" and (evolutionary) "continuity" senses.

INTERCONNECTEDNESS IN ECOSYSTEMS

Interconnections are the fabric of nature's well-being, ecotheologians argue. So long as these links are not damaged or severed, health will prevail and all beings will generally flourish. The ecological model of interrelationship "supports a holistic understanding of well-being: the health of nature and my health—as well as the health of other human beings—are interrelated";[8] biophilic mutuality and symbiosis are indications of nature's proper functioning, and these become "disrupted" by "human intervention" that brings death to biotic communities.[9]

As we have seen, ecotheologians regard ecosystemic interdependence (connections within ecosystems) in a pre-Darwinian and often Romantic light. By this I mean that the ecosystem as a whole is assumed to produce some higher, communal good—an ultimate, overarching end toward which all other proximate ends are directed. Nature's interdependence is understood as a benevolent force that courses through all forms of life, linking them to one another. Not only is there assumed to be some emergent, systemic good (something even Darwin, as well as Rolston, might agree with, in the sense that a species as a whole is perpetuated), but the well-being of *each member* of the community is also assumed to be sustained. Humans, in structuring their own communities on the blueprint of the ecological model, are likewise to respond to the bodily needs of each member in the community, fostering relationships of mutuality. Ecosystem interdependence is thus seen as a solution to a set of problems—the problem of suffering, power asymmetries, and domination that have attended our efforts to abstract ourselves from the web of life.

Organic interrelationship was a feature of Darwin's thinking as it was for earlier ecologists, but for Darwin struggle and competition were the very strands out of which the web of life was woven. In this sense interdependence is not so much a solution to strife and suffering as it is a *source* of it. In *The Origin of Species* Darwin reminds his readers that "the structure of every organic being is related to that of all the other organic beings with which it comes into competition for food or residence, or from which it has to escape, or on which it preys."[10] He was well aware of how easy it was, in pondering nature's apparent abundance and beauty, to forget the death and destruction that lay beneath the surface. "It is difficult to believe," Darwin confessed,

> in the dreadful but quiet war of organic beings going on in the peaceful woods and smiling fields. . . . We behold the face of nature bright with gladness . . . but we do not see or we forget that the birds which are idly singing round us mostly live on insects or seeds, and are thus constantly destroying life; or we forget how largely these songsters, or their eggs, or their nestlings, are destroyed by birds and beasts of prey.[11]

The Origin of Species contains repeated reminders, perhaps to Darwin himself as much as to his readers, that nothing is easier to forget than the universal struggle for existence. Yet if this struggle is not "thoroughly engrained in the mind, the whole economy of nature, with every fact on distribution, rarity, abundance, extinction, and variation will be dimly seen or quite misunderstood."[12]

Darwin explains that he uses the expression "struggle for existence" in different ways; modern Darwinians are still engaged in sorting out its various meanings and the different levels on which it operates. In one sense, Darwin argues, organisms will actually face off against one another for food in times of dearth. In these instances competition is most fierce between members of the *same* species because they depend upon the same kinds of food and other resources.[13] We can see, then, why Moltmann's concept of created niches and ready food supplies for each "kind" of organism is naive, since the struggle for existence can be most severe among members of the same species. The fact that species will have a common food supply and geographic range (as his concept of the ecology of space emphasizes) may lead to an increase, rather than a decrease, in struggle among species members. The second sense of struggling consists not of a direct competition between two animals but a struggle against the very conditions of life. A plant in extremely dry conditions, for example, is "said to struggle for life against the drought," Darwin observes, "though more properly it should be said to be dependent on the moisture."[14] Dependence and struggle go hand in hand because the resources upon which an organism depends are not always provided by nature. Food that is abundant one year may be extremely scarce in another. Nature fluctuates continually, Darwin argues, and modern ecology affirms this (without denying the existence of any order or regularity to nature). Darwin records the events of one winter in which the harsh conditions "destroyed four-fifths of the birds in my own grounds," adding that "this is a tremendous destruction, when we remember that ten per cent is an extraordinarily severe mortality from epidemics with man."[15] Nature simply cannot meet the individual, bodily needs of each member in the ecological "community." To say that an organism *depends* upon the presence of many other things in its environment is not to say that it will therefore *receive* what it depends upon. Nature's interdependence is no guarantee of well-being, as ecotheologians sometimes blithely assume. If we were consciously to emulate natural interdependence in a realistic, rather than a romantic, sense we would surely end up with *more* competition and struggle in our own society than we currently have. Gustafson recognizes this, arguing that any turn to nature for moral guidance that produces an ethic that is "indubitably beneficial to our species (or to other species)" must be constructed upon "an intolerably romantic basis."[16] Respect for natural processes—not naive reverence—is appropriate. A naturalized ethic cannot mean simply "following nature," conforming to its interdependence, in an uncritical fashion.

We have seen that ecotheologians often understand interdependence to be a benign condition natural to ecosystems, and thus they interpret suffering and strife to be alien, human-imposed problems. Human beings are to restore nature to its normal balance and, in doing so, restore themselves to nature. Restoration of both begins with the recognition of our own place within this larger web of life. Ironically, accentuating interdependence often puts humans in a position of having far more control over nature than the ecological model itself would seem to indicate. We are to alter nature in light of a perceived "ecological" objective, which in fact is merely a set of human expectations, and moral preferences, imposed upon the natural world. In other words, ecotheologians' concept of ecosystem interdependence not only misunderstands nature but misunderstands the extent to which humans are responsible for it.

A theocentric perspective, on the other hand, proposes that humans are participants in processes that are not completely under our control, and, as we will see in a moment, elements of land ethics affirm this understanding of human moral obligations. Humans as participants is a concept closely connected to recognition of nature's interdependence, but it means something different for Gustafson and land ethics than for ecotheologians. I want to argue that a proper interpretation of interdependence, to which we now turn, also corresponds to Darwin's view of natural processes.

INTERDEPENDENCE AND CONFLICT IN LAND ETHICS AND THEOCENTRIC ETHICS

We have looked at Gustafson's critique of Western religion's anthropocentric tendency to derive ethics entirely from human needs and perspectives. This critique is conceptually linked to a particular view of interdependency: our "interdependence qualifies our tendencies toward anthropocentrism," Gustafson argues.[17] A recognition of our dependence upon and interdependence with nature reinforces the fact that we are not the center of things, that we are sustained by an ordering that was created neither by us nor for us. The fact that we are not the center of things—that we cannot experience the whole, but only parts of it—entails a "consciousness of the ambiguities of the many relations of multiple values of things for each other in nature."[18] Interventions in natural processes, if they are to be justified at all, must entail a considered response to a particular set of details, and even our most justifiable actions will generate consequences that may not have been foreseen. In Gustafson's view, *participation* is the word that best describes—and circumscribes—our moral action.

The idea that we are not in control but are merely participants in the natural ordering has great theological import for Gustafson, as we have seen. A sense of dependence—a sense of the "divine"—entails respect for processes that both sustain and destroy life on earth.[19] Nature sets certain limits and provides certain possibilities for our moral actions. We look to the natural ordering as a partial basis for deciding what goods to pursue, but nature's systemic order does not necessarily provide simple answers. Our "choices have to take account of contexts; the values are always in relation to them."[20] Ethical choices involve us in a process of ranking values, of relating parts to wholes, individuals to communities, and the conflicts that inhere in those choices are never entirely resolved. Gustafson notes that some theologians and ethicists "have proposed that at the deepest level, the good of individual parts and the good of wholes to which they belong are harmonious." But this is seldom, if ever, the case: "In the conditions of finitude and the need to make particular choices, this [proposal] is false."[21] The inability to resolve conflict sometimes creates a longing, especially for religiously minded individuals, for a world in which all values *can* ultimately be brought into harmony, and benefits can be realized by all beings at once. "We might idealize a world in which the values in particular relations work harmoniously for the good of the whole," Gustafson argues. Moreover, "we might eschatologize this, looking toward some coming or eternal kingdom in which this harmony is realized."[22] I have argued that ecotheologians engage in precisely this sort of "eschatologizing" of nature rather than consenting to the sort of interdependence that exists in nature. Gustafson rejects this turn to an ultimate, ideal harmony, as does Rolston, and in this sense—as well as in the details of their interpretation of interdependence—their views of nature are compatible with a Darwinian view.

Like Darwin, Gustafson interprets interdependence as inseparable from conflict. Interdependence is not simply another term for harmony and community—it is not the antithesis of conflict, as ecotheology often understands it. In nature, Gustafson argues, we find "interdependence without equilibrium."[23] Disequilibrium and disturbance in nature can have a natural or a human origin, but neither source can be completely eradicated. If we accept this, we must also accept the idea that it is not the task of humans to promote harmony in our interventions in natural processes. In nature there is "no clear, overriding telos, or end, which unambiguously orders the priorities of nature and human participation in it so that one has a perfect moral justification for all human interventions."[24] He notes that nature's equilibrium is often incorrectly assumed to follow from its interdependence. This is problematic because "it introduces a norm of what the 'proper condition' of each and everything is"—an assumption that is too static.[25] Given this account of natural processes, the most appropriate role for humans to envision for themselves—indeed, the only role possible—is that of "participants." It is not our place to intervene in order to control nature or to prevent the destruction of life that occurs naturally, no matter how distasteful it may seem. Life is such in the world, Gustafson argues, that "some things have to be destroyed for the sake of other beings."[26]

This understanding of interdependence, which I take to be essentially Darwinian, is also highly compatible with the perspective of land ethics. This is perhaps not surprising, since both approaches take scientific data seriously. Yet the convergence of land ethics and theocentric ethics on the language of "participation" in natural processes as the appropriate description for human interventions is striking.

Before turning to the land ethics account of participation, however, I want to emphasize, again, that I am not arguing that land ethics is inherently a religious position but, rather, that it is compatible with a theocentric perspective. Gustafson stresses that the basic moral *response* toward nature that his perspective entails is not a formula for arriving at solutions to ethical problems: "The moral stance itself does not resolve environmental ethical issues."[27] Part of the reason that interdependence does not guide us to a clear or unambiguous ethic is that nature is constituted by both conflict and order (in this, Darwin—as well as many modern ecologists—agrees). Nature itself is morally ambiguous and its interdependence presents us with a multidimensionality of values. But for Gustafson there is another reason why interdependence does not provide moral answers, namely, because his understanding of ethics involves a distinction between a general moral framework and specific instances of moral decision making. A sense of dependence on and interdependence with the natural ordering is not taken to be a *solution* to problems or conflicts in nature, as it is for ecotheologians who move too quickly from description of nature to prescription. Rather, this basic sense constitutes a starting point for moral action, much as Leopold regards an "intuitive" respect and love for the land as necessary but not sufficient for developing a land ethic.

Gustafson distinguishes what he calls a "common sense ontology" (which acknowledges the *fact* of interdependence) from the more complex process of drawing ethical comparisons and making judgments. In other words, our dependence evokes awe and respect, but such responses (which may be manifested as either a religious or a secular/moral response) do not tell us, unambiguously, what we are to do. "Interdependence," he argues, "does not automatically issue in harmony of desired and desirable ends."[28] A moral stance, while important, is insufficient to answer the questions "Good for whom? Good for what?"[29] These questions must be answered in their specific contexts: "The particular choices are not determined by the stances."[30] Gustafson has a clearer understanding of the difference between scientific fact and moral norm than many ecotheologians. More important for the point I want to make here, his distinction between the general moral stance (dependence, awe, respect) and the specific choices that are made suggests that land ethics, as an approach that ranks values, is compatible with the general stance of a theocentric perspective.[31]

The questions that follow from a theocentric orientation—Good for what? Good for whom?—are the sort that land ethicists such as Rolston and Callicott attempt to answer, with reference to values that are not exclusively human centered. Indeed, as one scholar has noted, "Gustafson's thought is most similar to Rolston's in trying to affirm human value while recognizing how interactive relationships affect both human and non-human values and valuations." Both Gustafson and Rolston "refute narrowed, anthropocentric claims of traditional Western Christian and philosophical rationales."[32] As a method of discerning and prioritizing values within a general framework of respect for natural processes, land ethics coheres with a theocentric orientation. Respect for natural processes is a general stance in land ethics, within which competing claims "may be adjudicated and relative values and priorities assigned to the myriad components of the biotic community."[33] Whether or not land ethics is a theocentric perspective (i.e., whether or not the sense of dependence and respect for nature is a *religious* affectivity) is a matter of personal conviction. Although Gustafson considers this sense of dependence to be a sense of the divine, he acknowledges that one may have a strong moral response to nature without interpreting it religiously: "I have no serious quarrel," he writes, with authors who "have similar moral stances toward nature but are agnostics or declared atheists. . . . I cannot persuasively argue that their stances entail the acknowledgment of God or gods, nor do I wish even to try."[34]

A secular orientation may be generally compatible with a theocentric orientation without understanding *consent* to the powers that sustain us and all life as a religious movement or a leap of "faith." If, as I am arguing, land ethics is compatible with a theocentric perspective, then the two ought to understand the human role regarding interventions in nature in a similar fashion. This in fact is the case.

INTERDEPENDENCE AND PARTICIPATION IN LAND ETHICS

We have seen that the biotic community in land ethics does not mean the same thing as community in human ethics (or community as defined by ecotheologians). Such differences in orientation between land ethics and much of ecotheology may account for why Leopold's work has had relatively little impact in religious environmental ethics. Judith Scoville notes that while many ecotheologians "cite with approval Leopold's call to extend ethics to the natural world . . . most references to Leopold in theological literature are of the 'sound bite' variety: they make brief mention to illustrate or bolster a point but do not seriously engage his thought."[35] Specifically, Scoville argues that some theologians perceive a "fundamental conflict" between the ecocentrism of land ethics and theocentric grounding of (certain) Christian theologies, insofar as Leopold gives clear ethical priority to ecosystems, to biotic, rather than human, communities.[36] In fact, this is precisely the objection issued by ecotheologian Michael Northcott who dislikes land ethics' decentralization of humans, arguing that "the ecocentric ethical priority which the land ethic gives to the stability of ecosystems is not consistent with a Hebrew and Christian ethic, for it seems to locate moral value primarily in the biotic community of the land."[37] Another problem with Leopold's ethic from the standpoint of theology is that the value of *individual* (nonhuman) entities is relative to the larger biotic community, whereas many ecotheologians view ecosystems as subordinate to the needs of the individual members (human and nonhuman) of the community. In general, land ethics appears to some theologians to make the land itself, as Leopold defines it, into "god." This impression of a basic conflict between much of ecotheology and land ethics is rooted in a misconception that Leopold placed humans in a subordinate role to nature, according to Scoville. Thus she prefers to label Leopold's ethic "ecological" rather than "ecocentric" in order to deemphasize the apparent centrality of nonhuman nature in land ethics.

Scoville's perception that the ethical concerns of Christian theologians are often at odds with those of land ethics is correct. Her claim that land ethics affirms negativities in nature, and that "similar affirmation" is needed in ecotheology, is also on target. I am in complete agreement with her claim that Leopold's work "challenges us to make our environmental ethics fit the way the natural world actually works."[38] I take issue, however, with the argument that the conflict between ecotheology and land ethics comes down to a clash between theocentrism and ecocentrism: as I have argued, many ecotheologians *fail* to construct an ethic that is genuinely theocentric. Indeed, were they more successful in their efforts to develop a theocentric perspective, points of conflict with land ethics would decrease rather than increase.

Interdependence in land ethics is consistent with Darwin's understanding (the interconnections are themselves constructed out of complex chains of competition and struggle that involve death for individuals) and with theocentric ethics (natural processes do not necessarily guarantee benefits and desirable outcomes, nor can conflicts between individuals and communities be fully harmonized). Like Gustafson's theocentric ethic, land ethicists such as Callicott also stress that we are "participants" in processes we did not create and that do not always match human moral preferences and expectations. Land ethics aims at maintaining the integrity, stability, and beauty of the biotic community, without trying to bring this community in line with a human moral sensibilities.

Recall Leopold's description of nature as a pyramid, each level of which is sustained by a process of eating and being eaten. As Callicott has noted, land ethics accepts this structure, affirming "natural, biological laws, principles and limitations" rather than imposing human moral codes onto nature.[39] As in Gustafson's account of human participation in nature, land ethics consents to a natural ordering in which one being lives at the expense of another. This view of nature, in Callicott's words, invites us "to reaffirm our participation in nature by accepting life as it is given, without a sugar coating."[40] Participation in land ethics similarly implies moral action as well as moral limitations—what Gustafson refers to as the dual conditions of finitude and responsibility. Both approaches take seriously the conviction that "in the sphere of nature, our interactions frequently require greater conformity to the ordering that is present than they permit a new ordering or even a mastery of the ordering that exists."[41]

We have seen that Rolston's ethic particularly emphasizes the appropriateness of consenting to the natural ordering rather than trying to reform or redeem nature. In land ethics interventions in natural processes must be well justified and cautiously executed. "Wildlife," Rolston argues, "is life that can manage itself, even if human wildlife managers have to arrange so that this can happen."[42] Intervention, in other words, is warranted because humans have already, and often inappropriately and recklessly, intervened. Again, Gustafson interprets participation and intervention in a similar light: "Prior events and occasions have established both the limitations and the possibilities for our intentional interventions," including those "into nature."[43] The fact that humans are not the center of things—that we cannot experience the whole, but only parts of it, even while we endeavor to discern what is good for the whole—entails a "consciousness of the ambiguities of the many relations of multiple values of things for each other in nature," Gustafson cautions.[44] Consciousness of ambiguities, in turn, evokes (or ought to) "greater modesty" precisely because we "are not privileged to know with clarity and certainty just what proper relationships and activities are in the ordering of life in the world."[45] Or, as Leopold puts it, "the scientist . . . knows that the biotic mechanism is so complex that its workings may never be fully understood." For this reason, we must remember our role as a "member of a biotic team."[46] There is a kind of modesty to each of these views—an acceptance of the conditions of finitude and the restraints upon human interventions in the natural world, as well as an acknowledgment of the non-negotiable fact of suffering. For all of these reasons, a genuinely ecocentric ethic finds support in theocentric ethics. If ecotheologians wish to speak with scientific authority about ecosystem interdependence, they have much to learn from more scientifically informed arguments such as these.

As I suggested at the outset, there is a second type of interdependence that warrants closer scrutiny, for it is one that many environmentalists directly attribute to Darwin's work. We will now turn to interdependence as *interrelatedness* and see whether ecotheologians' invocation of this concept as a basis for ethics is consistent with an evolutionary account.

INTERDEPENDENCE AS EVOLUTIONARY CONTINUITY

Environmentalists repeatedly cite the scientific fact of an evolutionary continuum between humans and animals. We share a basic nature with all other living things. This fact in itself ought to tell us something important about our ethical relationship with other animals. But what exactly does it tell us?

We have seen examples of this "Darwinian" position in secular arguments opposing speciesism as an arbitrary, unwarranted discrimination akin to racism. The ecological model of Christian environmentalists also proposes this "scientifically" based justification for regarding other animals as worthy of similar moral consideration: "The ecological model assumes multiple relations with many others, who are related to the self analogously along a continuum."[47] The ecological, evolutionary model implies that "we are not and cannot possibly be utterly different from all the rest."[48] Among Christian environmentalists interrelatedness is sometimes, but not always, expressed in terms of a love ethic. Secular approaches may prefer language of equal rights or interests, but the search for "sameness" among humans and animals is rampant in environmental ethics.

McFague's ethic is the most explicit in its insistence on a "loving" approach rooted in biological sameness, where love entails following the praxis of Jesus in caring for the bodily needs of other beings. Love and liberation are related concepts since liberating other beings from oppressive circumstances follows the example of Jesus's ministry and fulfills their physical needs: "In a time of the earth's deterioration," she argues, "Jesus' radical, inclusive love ought to embrace the earth others who are suffering bodily oppression."[49] Birch and Cobb draw on Christian language of love and liberation—they join McFague in condemning the objectifying gaze that oppresses life—but they also invoke more secular-sounding "rights" terminology to link humans' and animals' claims to protection and enriched lives. On the whole, many environmentalists agree that evolutionary sameness implies moral "sameness"—other animals are subjects (McFague, Regan, Birch, and Cobb) or "persons" (Singer); discrimination against them is prejudicial (recall, however, that Birch and Cobb do discriminate between certain kinds of organisms. I will return to this point in my discussion of the parameters for an ethic of love).

We have already seen that the Darwinian notion of ecological interdependence (the first sort of interdependence) is founded on more conflict and disequilibrium than ecotheologians often realize. What does a Darwinian perspective have to say about this second sort of interdependence—the idea that all beings exist on a continuum and ought to be treated as "analogous" to us? Does biological science provide a basis for regarding nonhuman animals as equally worthy of moral consideration, as environmentalists often contend? Does it support an ethic of "loving" our animal kin? Richard Ryder has said that a speciesist bias against nonhuman animals denies the very logic of evolution—speciesists pretend that animals are not their kin and do not share many of their capacities, experiences, and desires, all evidence to the contrary. Darwin himself, it is argued, maintained that a difference of degree rather than kind existed between humans and other animals.

It is certainly true that Darwin devoted much of his life to cataloging the similarities—mental, moral, emotional, and physical—between humans and animals. He argued that nonhuman animals, "like man, manifestly feel pleasure and pain, happiness and misery."[50] Animals experience terror as we do, they exhibit traits such as "courage" and "timidity"; they may be good- or ill-tempered. They "not only love, but have the desire to be loved," Darwin maintained. In *The Descent of Man* he describes one baboon with "so capacious a heart" that she regularly adopted the orphaned young of other primate species as well as an assortment of kittens and puppies.[51] Moreover, as I have noted previously, Darwin contends that many animals experience "intellectual emotions"—states of mind involving not merely instinctive or immediate reactions but capacities for reflection, memory, and anticipation; among these animal states of mind he counts ennui, wonder, curiosity, and dread. Perhaps most distressing to Darwin was the suffering of animals who possessed these sophisticated mental and emotional capacities. He recounts a well-known story of a dog undergoing vivisection who "licked the hand of the operator" as if pleading for mercy; "this man," Darwin added, "unless he had a heart of stone, must have felt remorse to the last hour of his life."[52] It is no wonder that animal liberationists such as Singer and Regan see Darwin as a forerunner in their cause.

Despite his great interest in and concern for animals, however, Darwin declined to join the cause of antivivisection or to swear off a diet of meat. He did not encourage a program of "liberating" domesticated animals, nor did he promote a policy of reducing the suffering of animals in the wild.[53] Did Darwin deny, at least in practice, the logic of his own theory?

Darwin himself is not much help on the question whether evolutionary continuity implies moral continuity or even an ethic of love (although he alludes to animals' capacity for loving and being loved). For one thing, he was deeply ambivalent about many issues pertaining to animal welfare. Furthermore, scientists in Darwin's time had no access to the kinds of genetic arguments that are often invoked to bolster moral claims. He knew that animals were our kin but was not able to quantify such relationships in the way that biologists commonly do now (as when we are presented with the now commonly cited fact that chimps share almost 99 percent of our genes).

Modern Darwinians have taken up the task of clarifying how our evolutionary nature informs our ethical relationships with nonhuman animals and nature as a whole. Let us look at two authors who propose theories regarding the biological basis for our responses to nature in order to understand how evolutionary continuity might contribute to moral consideration of all life-forms. We have already encountered one of these authors, Mary Midgley, in chapter 4. The other theory, and the one that we will consider first, comes from sociobiologist E. O. Wilson.

E. O. WILSON'S BIOPHILIA HYPOTHESIS

In *The Biophilia Hypothesis* E. O. Wilson attempts to explain the evolutionary and psychological foundations for human responses to animals and nature as a whole. In Wilson's account we see elements of what critics such as Gould have labeled an "adaptationist" preoccupation, as well as tendencies toward genetic determinism in explaining human behavior. Biophilia refers to the "innate tendency to focus on life and lifelike processes," where "innate" means originating in our evolutionary past. Wilson's basic argument is this: over a long period of evolution the human species acquired a set of "learning rules" that enabled us to survive in different kinds of natural environments. These learning rules were reinforced both genetically and culturally, through a process known as "gene-culture" coeveolution: "a certain genotype makes a behavioral response more likely, the response enhances survival and reproductive fitness, the genotype consequently spreads through the population, and the behavioral response becomes more frequent."[54] Genetically influenced behaviors are further augmented and perpetuated by cultural developments, such as myths, narratives, dreams, and even religious beliefs, that reinforce the innate predispositions.

It is easy to grasp how this process of gene-culture coevolution might take place when we consider natural phenomena that are dangerous and usually generate an aversive response. An innate, negative response to snakes, for instance (an example frequently cited by proponents of the theory), would incline a person or animal to avoid them whenever possible. Such avoidance has obvious survival value and, in human culture, is further reinforced by cultural artifacts such as myths that vilify snakes and recount our ancient enmity with them. (Genesis contains a familiar example and other cultures have their own such tales, Wilson argues.) Predators, parasites, and venomous insects, as well as natural phenomena such as lightning, have likewise been threatening elements in our environment throughout evolutionary time, and, as we might expect, many people have aversive responses to them. Furthermore, it appears that aversive reactions to such things occur even upon a first encounter (as with children who see a snake for the first time), and even when contact is not direct (seeing a picture of a snake or watching one on a nature program). Aversive responses are formed with little or no negative reinforcement.

The prevalence across cultures—and across species—of such bio*phobic* responses, as they are called, is taken as evidence supporting the hypothesis. Primates such as old world monkeys and apes display a similar "natural fear" of snakes and serpents. Other common biophobic responses include fear of enclosed spaces (such as caves), heights, and running water—natural phenomena that can easily be understood to have posed a direct threat to a species that evolved in an intimate relationship to the natural world.

Many positive, "biophilic" responses to nature are also widespread, and these, of course, are the responses that might foster an environmental ethic. Flowers, for instance, seem to be universally attractive to humans, as they are to many animals and insects as well. Wilson and others point out that flowers signal the presence of food resources (flowering plants bear fruit). Studies have also shown that many people prefer trees with a particular shape, such as those with low-hanging branches and a large canopy, presumably because such trees provide good shelter or are easily climbed in escaping a predator.[55] At the same time, most people have negative responses to barren, treeless landscapes such as deserts, environments that would have provided our ancestors with few resources (food, water) and little or no place to take shelter from predators or storms. What first appear to be individual aesthetic preferences may turn out to have a specieswide, functional origin in our evolution, proponents of biophilia contend. Aesthetic responses may express ancient affinity for places ensuring "greater likelihood of food, safety, and security associated with human evolutionary experience."[56]

Evidence for biophilia can also be found, some biologists argue, in human responses to animals who are our close evolutionary kin. This aspect of the biophilia hypothesis has direct bearing on the question whether a biological continuum implies a moral continuum. A sense of kinship is "stamped by a common genetic code," Wilson argues.[57] Given that social bonds are formed with animals who share our genes (the strongest bonds usually being formed with those in our own species), and since we share some percentage of our genes with *all* living things, a biophilic response toward all life may be genetically reinforced. Special interest and concern for organisms who share many of our genes is also the basic idea behind the selfish gene theory, about which I will say more shortly.

Genetic arguments seem to lend support to the claims of ecotheologians and other environmental ethicists who maintain that the fact of evolutionary kinship can—or should—produce an ethic of love or care for beings with whom we share the world. If it is true that positive responses to animals correspond with degrees of evolutionary relatedness, then Birch and Cobb's richness of experience scale may have a biophilic component to it. That is, they seem to give preference to animals who also happen to be our closest kin, such as primates and mammals as a whole. Yet such preferences can be inappropriate as an environmental ethic, as we have seen, promoting a homogenization of life rather than an appreciation of biodiversity. It appears then that biophilia, like a richness of experience criterion, may similarly go awry, causing us to be concerned primarily with the individual well-being of certain animals closely related to us. Furthermore, these (selectively) positive responses intermingle with blatantly negative ones, according to the biophilia hypothesis: we may be genetically related to spiders or scorpions, but many people still prefer to avoid any contact with them and might even wish for their total eradication. Genetic kinship with them is not strong enough, it seems, to counteract the basic biophobic response.

It would appear then that we encounter certain difficulties in attempting to derive from biophilia an ethic of love for the myriad life-forms to whom we are (more or less) related, or even a conservation ethic.[58] Setting aside the empirical question whether there is sufficient evidence to support the hypothesis (and particularly, the "unabashedly Neo-Darwinian" assertion that genes direct our behavior),[59] there are other problems with biophilia as the basis for a broad environmental ethic. Biophobic responses seem to interfere with our ability to extend moral consideration along the entire evolutionary continuum of organisms; biophilia only helps in some cases. It may, in fact exclude altogether certain species from ethical consideration, including the numerous slithering and multilegged creatures that make their homes in notoriously threatened tropical rainforests. Biophilia in and of itself can generate a variety of responses to nature, some positive, some negative, "from attraction to aversion, from awe to indifference, from peacefulness to fear-driven anxiety."[60] It is difficult to translate a response of terror into one of protection. Again, we see that the fact of interdependence (as genetic interrelatedness) does not in itself produce the desired environmental ethic.

Wilson would deny that biophilia generates obstacles to a broad ethic for all life. On the contrary, he argues that it may be our best hope for developing a conservation ethic. He notes that humans evolved in response to an environment that was biologically quite diverse; whether or not we can survive in a vastly simplified natural world remains to be seen. We may imperil our own species by radically altering the natural context in which we developed and thus conservation is clearly in our own interest as well as the interest of other beings. This, he concedes, is an anthropocentric argument for preserving biodiversity—one of several anthropocentric rationales that Wilson offers, such as the potential for wild species to provide crops, pharmaceuticals, and other resources for us. He refers to this argument as a "richly textured," enlightened form of anthropocentrism. The entire logic of the biophilia hypothesis is, in a sense, anthropocentric, insofar as what we value in nature is directly connected to its function for us and our survival (here we see elements of the adaptationism to which Gould et al. take exception). The anthropocentrism is mitigated only by the fact that utility value may not necessarily be something of which we are conscious when we respond to nature; our response may simply be a vestige from our evolutionary past, when survival in a dangerous environment was our primary concern.

However, Wilson denies that biophilia rests solely upon recognition of the "utilitarian potential of wild species," and he remains optimistic that his theory can contribute to an ethic encompassing a larger set of values in nature.[61] How might biophilia contribute to an ethic that embraces all lifeforms and values biodiversity? When extended further, to include a broad spectrum of beings, the limitations of biophilia become apparent. It is not clear, for instance, that there is anything inherently "rewarding" about natural diversity (in the sense that it would confer some adaptive advantage). It may well be that, far from being pleasant, a great amount of diversity is "overwhelming" and "innately less appealing" than a more uniform environment, the vicissitudes of which would have taken our ancestors less time to learn and adjust to when their survival was at stake.[62] Perhaps our dramatic simplification of our planet's original biodiversity reflects an evolutionary *aversion* to extreme diversity! It is also not obvious that humans should prefer wild animals to domesticated ones, so long as we share genetic material with them. Our innate attraction to certain species might easily be satisfied by maintaining a representative sample of them (or the most interesting ones) in zoos and parks. Biophilic responses do not seem to be discerning enough to guide us toward conservation of wild animals and their natural habitats. Contrary to Wilson's claim, interdependence and interrelatedness (past or present) do not seem to support a robust ethic.

Wilson seems to be aware that biophilia might allow some species to fall through the cracks. He invokes the "aesthetic and spiritual value" of *all* life's diversity but acknowledges that the biophilia hypothesis *at present* cannot easily explain or support such values. "The evolutionary logic is still relatively new and poorly explored," he concedes, and "therein lies the challenge to scientists and other scholars."[63] In the meantime, until the challenge is met of accounting for spiritual responses by means of the biophilia hypothesis, these responses remain useful in environmental conservation. Wilson's conviction that a broad, evolutionarily based conservation ethic will one day be developed—one that can account even for spiritual values—expresses a scientist's inordinate faith in his own enterprise. Moreover, the spiritual value of nature to which he refers is specifically the "service to the *human* spirit" that nature can render.[64] Richly textured or not, such value remains very narrowly construed.

Mary Midgley's arguments are similar in some respects to Wilson's but, in my view, they are more promising as a basis for an environmental ethic that considers intraspecies bonding as an important factor. Midgley acknowledges that her claims about bonds between humans and other animals imply that there are limits to how far we can extend our social bonds beyond our own species (something Wilson does not readily admit). She also offers an account of human "nature," based on inferences from biology, and her argument here is more nuanced than Wilson's. Human nature is much more flexible (less deterministic) in Midgley's account, and this flexibility implies that we can supplement, as well as restrain, our basic biological predispositions toward other forms of life. In other words, a biological basis for our response to other beings is relevant to Midgley's account, but it is not the only important factor in developing an environmental ethic.

MARY MIDGLEY ON SPECIES BONDS AND NEOTENY

Midgley agrees with Wilson that humans, like many other species, are naturally "bond-forming" creatures, and she too regards the fact of our interrelationship with them as morally significant.[65] Along with other animals, we tend to express a preference for members of our own species, but this does not bar us from forming bonds with other animals as well. The species boundary is real (hence, as we have seen, the parallel of speciesism with racism does not hold). Unlike racial preference, natural preference for one's own species is not merely a "product of culture."[66] For example, in some species these bonds are established and reinforced by means of imprinting with a parent. Imprinting can become somewhat more generalized, allowing an organism to bond with other members of its species but usually not with those beyond the species barrier. While a biological reality, species preference is far from absolute for many animals. Some species, including humans, will fail to recognize members of their own species as such (a phenomenon Midgley refers to as pseudo speciation), while including other animal species as part of their community. Animals that live in social groups requiring a certain amount of cohesiveness and cooperation can more readily extend social bonds to include other species than can solitary animals.

Humans seem to have a particularly strong tendency to reach out to other animals, but some nonhuman animals express similar behavior, deliberately choosing the company of others—including humans—that lie beyond the species barrier. Biologist John Terborgh has described cases of what he calls "homophilia" ("a friendly feeling toward humans") in wild animals.[67] One of his examples provides a good illustration of Midgley's argument that some animals can extend bonds to other species. He offers a fascinating first-hand account of an injured peccary (a piglike animal that normally lives in herds) who sought the company of humans at a research station in the Amazon. The sick animal stood near a heavily trafficked path in the woods, stepping aside as researchers passed by. No one at the station fed the animal or engaged in any direct physical contact with it during its extended visit, but their proximity seemed to provide some comfort to the peccary. "Had the animal wanted to distance itself from further contact with humans, it could easily have done so," Terborgh writes, but "it remained in the middle of Trail 1 by day, and by night, we discovered, it bedded down just a few feet from an investigator's tent."[68] Normally a social animal that seeks the "safety in numbers" that a herd offers against predators such as large cats, the peccary apparently sensed that being near a group of humans would provide similar protection in its vulnerable condition.

> Our peccary must have decided that the risk of consorting with humans was less than the one it faced by remaining alone in the forest. Perhaps it had noticed that the jaguar was seldom in the vicinity of the station. Whatever

> its reasoning, the *huangana* was right. Its vigor and agility steadily improved until, one day, a herd of its species crossed Trail 1 and our peccary was gone.[69]

Interestingly, the peccary appears to have regarded its human companions as having a largely utilitarian value, as a "foil" against its enemies, as Terborgh puts it; humans apparently are not alone in turning to other species for the benefits they can provide.[70] Recognition of the use value of another species may be one of the few starting points for relationships between different species. But the more fundamental point here is that animals who are capable of forming social bonds with members of their own species can often come to regard other species as nonthreatening members of their own group. Whether or not the peccary exhibited genuine affection toward its human companions, its behavior was impressive to the biologists. Having been taught that wild animals instinctively avoid humans, Terborgh writes, "I was pleasantly surprised to learn that animals can occasionally overcome their inhibitions and see us as benign."[71]

Terborgh's story illustrates Midgley's claim that expanded social bonds are common among many species, human and nonhuman alike, and *might* reinforce a better human-animal relationship. However, Midgley is clear that such instinctive predispositions do not prescribe behavior in a deterministic sense, particularly where humans are concerned. "People have been strangely determined to take genetic and social explanations as *alternatives* instead of using them to complete each other," she observes.[72] Her own argument eschews genetic determinism and reductionistic approaches to human and animal behavior, relying instead on what she refers to as "open" and "closed" instincts. Closed instincts dictate behavior in specific detail; the nest-building patterns of birds are one example of fixed behavior. The behavior pattern is not learned and is often performed even when an animal is raised in isolation from members of its own species. Open instincts are not fixed in this way and do involve some degree of learning—they provide a general directive for behavior but there is a "gap left for experience."[73] Open instincts therefore permit a *range* of behaviors, some variety of expression, within a general tendency: "hunting, tree-climbing, washing, singing, or caring for the young" are general behaviors, within which individual patterns and methods may emerge. Learning enhances the effectiveness of a behavioral strategy, and the amount of learning that takes place may vary from one animal to the next.

Many behaviors, therefore, are both innate and learned.[74] Open and closed instincts occur along a continuum, with degrees of "openness" and "closedness," Midgley argues, and there are probably no completely closed instincts. Certainly, human behavior does not originate with, and cannot be explained in terms of, closed instincts.

The openness of human social instincts makes it much more possible to develop bonds with many other kinds of animals. As Midgley notes, such practices as owning pets and training animals would be far more difficult to initiate, and perhaps even impossible, if this were not true. "It is one of the special powers and graces of our own species not to ignore others, but to draw in, domesticate and live with a great variety of other creatures."[75] The animals easiest to domesticate are those that, like the peccary, live in herds or packs and possess strong social bonds within their own species. Dogs, for example, respond readily to social signals from other species once they begin to learn how those signals operate. In this sense there is nothing *inherently* unnatural or oppressive about domestication: its origins may lie in mutually beneficial arrangements between species with strong social instincts, both of which have something to gain from the relationship.[76] As with any relationship, of course, domestication may *become* oppressive (if that is the right term) when relationships with animals become abusive and exploitative.

The ability and apparent willingness of humans to form bonds with species outside of our own may also have something to do with the fact that humans are a particularly "neotenous" species. Neotenous species are those that tend to retain juvenile characteristics into adulthood. Such characteristics may be physical (large eyes and a rounded cranium are juvenile traits) or mental and emotional. One important trait that humans appear to retain into adulthood is intense curiosity about their environments. Many species exhibit this trait as juveniles because curiosity and learning are extremely important for developing survival skills. "The human baby makes a beeline for the cat," Midgley notes. "The cat, if it is a kitten, returns the compliment with particular fervour."[77] As they mature, however, the behavior of many nonhuman species often becomes less flexible and more programmed. Engaging in play behavior teaches an animal a great deal about its environment, what resources it provides and what dangers it may hold. Once this information is gathered, the learning curve levels off and the initial inquisitiveness of the juvenile animal loses some of its intensity. So humans are not exceptional in displaying strong curiosity and play behavior, but, Midgley argues, we tend to retain such traits into adulthood. "The real corner-stones of these prolonged and intensified infantile faculties, however, probably do not lie in play itself, but in the sympathy and curiosity which underlie it."[78]

What do these biological arguments regarding interrelationship and innate responses imply about ethics? Neoteny and natural social bonds may be primary means by which we reach out to other species and take an ongoing interest in them. Whether or not such a tendency to respond to animals develops into a genuine ethic of sympathy or even love is less clear. Scientists, for example, would seem to exhibit a high degree of neoteny in their insatiable curiosity about the natural world, yet many have no serious qualms about harming or killing organisms in the course of satisfying curiosity and pursuing attraction to other life-forms. Consider the paradoxical attitude toward other life-forms expressed in one scientist's account of a good day's work in the field:

> I turned to watch some huge-eyed ants with the formidable name *Gigantiops destructor*. When I gave one of the foraging workers a freshly killed termite, it ran off in a straight line across the forest floor. Thirty feet away it vanished into a small hollow tree branch that was partly covered by decaying leaves. Inside the central cavity I found a dozen workers and their mother queen—one of the first colonies of this unusual insect ever recorded. All in all, the excursion had been more productive than average. Like a prospector obsessed with ore samples, hoping for gold, I gathered a few more promising specimens in vials of ethyl alcohol and headed home.[79]

The scientist who wrote this account is E. O. Wilson. Wilson describes his pursuit of knowledge as a sort of "magellanic voyage" that engages "the things close to the human heart and spirit."[80] Genuine reverence for nature is here combined with detached, meticulous collecting and killing of "specimens." Yet both revering and killing are expressions of the basic "biophilic instinct" that attracts the scientist to nature. Clearly, the attraction itself does not generate an unambiguous ethic toward other life-forms. Yet, as we have seen, Wilson wants to maintain that our biophilic responses can nevertheless be put in the service of a conservation ethic that can also be defended on other, rational, grounds. "Instinct is in this rare instance aligned with reason."[81]

In and of itself, curiosity and attraction to nature, like the biophilic instinct Wilson describes, can produce good or bad results. Possessing an ability to sympathize with other creatures, to recognize their suffering, is also no guarantee that those creatures will be well-treated, since even exploitation requires sympathy, Midgley points out. A person who abuses an animal would not bother doing so if he or she did not believe the animal could feel pain similar to humans. Cruelty to animals occurs in spite of—and perhaps because of—knowledge that they suffer (one more reason an ethic founded upon the common capacity in humans and animals to feel pain can be inadequate). The human tendency to feel sympathy and curiosity toward other species is important, but these evolutionary factors do not sufficiently guide our behavior.

We have seen that our ability to form social bonds with other animals (and their ability to respond in kind) may account for the development of domestication and pet keeping, according to Midgley. Once certain species became domesticated (particularly dogs and cats), relationships of bonding and even love became the norm. But what does our evolutionary heritage tell us about what human relationships are, or ought to be, with *wild* species of animals? Here it seems we have to supplement our natural bonding inclinations with other kinds of considerations. Whereas we may be free to indulge our feelings of sympathy and affection for domestic animals and pets, we need to think more carefully about how we express these feelings toward animals in the wild. We do not possess wild animals in the way we may possess pets; usually we are not able to touch them or interact with them as we do pets. (The popularity of petting zoos probably reflects this desire to engage normally elusive animals as though they were pets.)

I have argued that our tendency to be attracted to other animals does not necessarily tell us how we should treat them. Wild animals are a case in point, because even when human curiosity and sympathy toward them are manifested in the most positive light (that is, when they do not breed destructive inquisitiveness or exploitation) they may still produce inappropriate responses toward animals in the wild. It seems that where wild animals are concerned our natural feelings and our capacities to bond with other species must be channeled into some other kind of ethic, something less possessive and more detached, something less likely to interfere with their lives and their natural trajectories. In the case of the injured peccary cited above, for instance, the researchers kept their distance, attempting neither to feed nor heal it, saddened though they were that it "seemed to be in decline."[82] They did so, moreover, despite their obvious interest in, and perhaps affectionate feeling for, this animal—feelings reflected in the biologist's characterization of the encounter as "deeply touching" and his reference to the animal more than once as "*our* peccary."[83] An ethic of love toward wild nature and animals, if this can be an appropriate ethic at all, must be of a different sort than that which we extend to domestic animals.

What, in the final analysis, do Midgley's arguments and Wilson's theory of biophilia tell us about human nature and our relationship with other living things? Does evolutionary kinship lead to ethical consideration for all? Can evolutionary factors support an ethic of *love* toward all nature? The answer would seem to be a frustrating "yes and no." On the affirmative side such biological knowledge seems to imply that cultivating a feeling of love toward nature is not at all far-fetched as the basis of an ethic; indeed, we may not be able to avoid such a response. Yet, as the flip side of biophilia—biophobia—suggests, there also may be aspects of nature that are difficult for us to respond to in a positive way. Of course, we might try, as McFague would urge us to do, to develop a radical love of *all* creatures, just as Christians are urged to love their enemies, and attempt by some force of will or change of heart to overcome our aversion. But it seems better to consider the possibility that different kinds of animals ought to be treated differently, rather than trying to include all living things under the umbrella of a Christian love ethic, particularly one modeled upon human-human relationships. An ethic such as McFague's is broad but not well defined: the Christian imperative to love is not qualified by other considerations about nature—our own nature and the nature of the world we live in.

Yet a purely scientific account of our relationship to nature and animals may also be insufficient as the basis for an environmental ethic. If an ethic such as McFague's is too broad, the perspective of biophilia seems too narrow. At best, humans are relegated to narrow ranges of response toward nature that can have either positive or negative effects. At worst, we have been genetically engineered over millennia to respond negatively to a potentially very broad spectrum of organisms. If a Christian ethic of love does not discriminate enough between types of organisms, Wilson's seems to discriminate too much.

The biological fact of our interdependence with other life-forms, including the genetic basis for kinship, does not issue in a straightforward, natural ethic. The relationship between the facts and the ethic is more complex than a simple conformity to nature. I will return shortly to this complex relationship, in the course of defending a qualified love ethic. Before doing so, it is worth dwelling on these debates about human nature a while longer. Midgley makes an important point about our ability to construct an ethic that takes our "nature" into account but does not simply *default* to our nature (much less our genes) as some sort of infallible guide to conduct. Her basic argument is echoed by Rolston's proposal that humans occupy the world intellectually and morally as well as biologically. Rolston and Midgley's views on this subject are relevant to the question whether it is in *our* nature to love nature.

LOVE IN LIGHT OF *HUMAN* NATURE

Like Wilson, Midgley suggests that our responses to nature and animals contain an evolutionary component; the responses themselves may have positive or negative ramifications. Wilson maintains a strong link between our innate, evolutionary "wiring" and our behavior, and this stronger link makes for a more limited ethic. We are less free to detach ourselves, ethically, from our nature, in his view. Our genes keep us on a "leash," Wilson argues. Granted, he does acknowledge our ability to join reason with instinct, implying that perhaps we can stand back from our immediate, innate responses, or subject them (to some extent) to moral deliberation. He also suggests that cultural development (gene-culture coevolution) intersects with our genetic propensities, but he seems convinced that culture primarily reinforces our inborn propensities; reason is closely connected to self-interest, Wilson believes (reason evolved as a tool for our survival): "The more the mind is fathomed in its own right, as an organ of survival, the greater will be the reverence for life for purely rational reasons."[84] The "rational reasons" he has in mind are, as we have seen, anthropocentric ones—our survival depends upon conservation of natural resources. Reason is capable of grasping what our selfish genes already know. "The only way to make a conservation ethic work is to ground it in ultimately selfish reasoning"; our moral predicament calls for "biological realism."[85]

Perhaps Wilson's approach to the environment would be effective if each person on earth could be convinced that preserving every species, *every single piece* of the biodiversity puzzle, is in his or her interest. But as we have seen, the theory of biophilia itself suggests that this is not possible: there are organisms toward which we are instinctively indifferent and others of which we may be deathly afraid. Evolutionary survival has dictated that we respond positively only to those things that aid in our survival. Rolston raises a similar point when he wonders whether it is possible to derive a "love of all forms of life" from a theory of "selfish genes."[86] He notes that Wilson's argument encounters difficulties when it attempts to explain how altruism confers any survival/reproductive fitness value when such altruism is extended broadly to nature. In general, selfish gene theory explains apparently altruistic or cooperative behavior in terms of reciprocity and kin selection. When an organism takes on risks to itself in order to protect the life of another, it may do so for one of two reasons, both of which can supposedly be explained from the "selfish gene's-eye view" as Rolston puts it. Either the beneficiary of altruism is kin to the (pseudo-)altruistic one, in which case a considerable percentage of the latter's *own* genes are preserved even if it should die, *or* the altruistic organism reaps benefits in the form of reciprocal acts that enhance its own survival (e.g., food sharing). The second explanation accounts for animals' engaging in a "tit for tat" strategy (as selfish gene theorists call it), regardless of whether genes are shared between them.

Certainly it is difficult to understand how a *holistic* environmental ethic that aims at the preservation of an entire ecosystem could be built upon selfish genes: how, for instance, would one derive a land ethic from this theory, where an ecosystem is the entity to be "loved," preserved, and given priority over the individual gene-bearing members? These individual parts (individual organisms) are less important than the whole in land ethics. Indeed, their lives may be sacrificed for the good of this whole. Moreover, those organisms whose lives are sacrificed may be more closely related to us than is the species that will be preserved (as when a sentient goat species is culled in order to save plants). The plant species cannot even reciprocate our altruistic action, and thus our interest in preserving it is not easily explained.

So why should we (or our selfish genes) care? Wilson maintains that caring for the whole is in our self-interest, but, even if he is right, this kind of self-interest does not seem to be the same thing as *selfishness*. Ultimately, he urges us to adopt an ethic toward nature that embraces its totality, a vision "he has himself embraced but cannot quite reach on the basis of his theory."[87] In contrast to Wilson's theory, Rolston proposes that humans occupy their worlds "ethically and cognitively," not *just* biologically or genetically.[88] We have the ability to "see further" than other species do, to recognize that our actions have consequences that are wide-ranging—even global—and to try to act in ways so as to minimize negative effects on other life-forms and the environment. While all life may be biologically related, only humans have the ability to apprehend this information and to think about what it means. "Humans, alone on this planet, can realize that they are kindred with all." We have the power of "overseeing," Rolston states; humans are "worldviewers."[89] The *knowledge* that we are kin with other beings may be more significant than the kinship itself. We are in a unique position to do something with this knowledge.

The recognition that we inhabit a unique position is not the same as saying that we are completely separate from animals or that we have a divinely ordained role on earth, because such statements about humans downplay the evolutionary origins and long evolutionary history that have made our power of "overseeing" possible.[90] Nor does Rolston's interpretation of humans' role in nature simply translate into a belief that we ought to care for everything else in order to "save" ourselves. Our unique position can suggest something different, a "switch to a biocentric conviction arising from a love of life beyond self-love."[91]

What does Rolston mean by a love of life beyond self-love? Wilson might argue that he wants to protect biodiversity for ultimately selfish reasons, such as recreation, or that his apparent altruism is really a form of deceptive beneficence so that others will "laud" him, return his altruism, or perhaps assist his offspring. But surely, Rolston writes, it is better to say that "the 'self' has been elevated into genuine morality, where it can detect values outside itself, and come to embrace these values in freedom and love because it is right to do so. This is not naturalized ethics in the reductionist sense; it is naturalized ethics in the comprehensive sense."[92]

Rolston's phrase here—*naturalized ethics in the comprehensive sense*—gets at the heart of what constitutes an appropriate understanding of nature, both our own and the natural environment. I would argue this understanding of the intersection of ethics and nature is also embedded in Gustafson's theocentric ethics as well as Midgley's account of evolution and morality. Human beings seem to have a basic response to nature that often involves feelings of love, respect, or awe. But this response must be subjected to scrutiny because it does not necessarily tell us what we are to do and may lead us to respond in ways that are inappropriate in a given context. Gustafson, Midgley, and Rolston suggest that we are able to stand back from our "innate" or prereflexive responses and consider whether or not they are appropriate. Part of our "nature" as products of an evolutionary past may be reflected in these responses, but it is also in our nature to reflect *upon* these reactions. In responding to nature and animals with an ethic of love, we need to deliberate about it, contextualize it, and qualify it.

Perhaps an ethic of love can manage to be both broad and discriminating—i.e., one that includes a wide spectrum of nature without failing to draw morally relevant distinctions between different forms of nonhuman life. As I noted above, the basic, positive response that we (or some of us) have toward nature and animals does not in itself constitute an ethic. The specifics of our ethical obligations toward other life-forms have to be worked out with regard to the details of particular cases and particular types of beings. As I understand it, this is essentially Gustafson's point when he argues that a response (respect, awe, gratitude) toward nature does not *tell us what to do*. As participants in natural patterns and processes, moral action requires a process of discernment of values that may begin with—but does not end with—a general response, a moral stance regarding the natural world. Moreover, as I argued in the previous chapter, I think there is a similar, basic moral response underlying the arguments of ecotheologians: the desire to respond to the suffering of other beings, a basic predisposition to love or care for them, and a sincere hope for their deliverance from pain all reflect an intuitive, moral stance toward other forms of life. This response in itself is not necessarily wrong, but it requires a more critical interpretation. As we have seen, this intuitive response is insufficient because (unlike land ethics) it does not take account of the contexts in which our moral obligations arise (evolutionary, ecological contexts) and it assumes a role for humans that is too controlling. From a *theological* standpoint this basic response can be criticized in that it may generate a desire to intervene, to "save," when such action is inappropriate. This kind of ethic, as Gustafson argues, entails a rejection of the given ordering, a religious disposition of ingratitude, and a refusal to consent to the limitations that this ordering establishes.

Upon further examination, a *scientific* perspective also suggests that a general response to other life-forms does not in itself constitute an appropriate ethic. For one thing, as Rolston and Callicott's arguments stress, environmentalists' desire to prevent suffering of sentient beings can disrupt natural processes that ought to be respected, and it may inappropriately favor those beings who most closely resemble humans. Second, as Midgley and Wilson's theories suggest, the basic response to other life-forms (assuming such genetic responses exist) is itself morally ambiguous: these responses range from the extremely positive (love, affection, awe) to the extremely negative (dread, revulsion, terror). Even what appear to be positive responses such as curiosity, attraction, and awe are, at best, morally neutral, since they can generate behaviors ranging from protection to vivisection of other organisms. A general, positive reaction of love or affection may also be misapplied, as when we desire to treat a wild animal as though it were our pet dog or cat.

The point, then, is that it appears that humans have a general, prereflexive, and often (though not always) positive inclination toward many other life-forms.[93] Gustafson understands this as a religious response to the ordering that sustains life, but he also recognizes that the response itself is rather common, expressed by nonreligious persons as a sense of the sublime (rather than the divine) or a kind of natural piety (rather than genuine religious piety). Indeed, he cites Midgley's work as an example of a nonreligious moral stance toward nature that resembles his own. Such responses to nature, he seems to believe, are widespread. Many scientists and secular ethicists apparently agree that humans have this general capacity. Leopold himself spoke of love and respect for nature as a dispositional prerequisite for developing a land ethic. "It is inconceivable to me," he wrote, "that an ethical relation to land can exist without love, respect, and admiration for the land."[94] Thus the land ethic is, in a sense, an expression of love for the community of other life-forms. "This sounds so simple," Leopold acknowledges. But upon closer inspection, love appears to have gotten us nowhere in terms of providing significant concern or protection of nature. "Do we not already sing our love for and obligation to the land of the free and the home of the brave?" Leopold asks. But, he adds,

> just what and whom do we love? Certainly not the soil, which we are sending helter-skelter downriver. Certainly not the waters, which we assume have no function except to turn turbines, float barges, and carry off sewage. Certainly not the plants, of which we exterminate whole communities without batting an eye. Certainly not the animals, of which we have already extirpated many of the largest and most beautiful species. (204)

Love and respect for the land is necessary for the development of land ethics, but it is not sufficient. "Man always kills the thing he loves," he laments, "and so we the pioneers have killed our wilderness" (148). Even a love that is expressed as protection of other beings can be problematic. We have seen that Leopold's sense of love does *not* dictate that we intervene to protect predators from their prey, thereby destroying the biotic pyramid he describes. He is also sharply critical of what he considers a "primitive" ethic of valuing nature to the extent that it can be "possessed" and experienced firsthand. The "trophy-hunting" recreationist loves the wilderness only as something that "he must possess, invade, appropriate" (176). "Hence," Leopold claims, "the wilderness that he cannot personally see has no value to him" (177).

Leopold as well as ecotheologians use the language of love, at least from time to time, to describe our ethical response. Midgley and Wilson also propose biologically based accounts of love and affection (bonding, caring) for other creatures. The idea of a love ethic does not adequately describe the basic orientation of all the environmentalists whose arguments we have examined. Yet it is the only term broad enough to capture the various ways in which our moral response to nature is characterized by these authors.[95] The question is, What are we to do with love for nature?

Again, Midgley's arguments are helpful in answering this question. Her account of human curiosity and species bonding, like Wilson's biophilia, does not in itself guarantee that an ethic of love will flow from our "nature." Yet when we combine her argument about humans' ability to form social bonds with her understanding of how we negotiate conflicts and deliberate morally, we begin to see that humans are in a unique position to extend moral consideration widely. Let us look more closely at Midgley's description of human moral deliberation.

Midgley proposes a model of moral reflection that is far more subtle and complex than many biologically based accounts of human morality. When one talks about the "nature" of a species, what this really means, as we have seen, is "a certain range of powers and tendencies, a repertoire, inherited and forming a fairly firm characteristic pattern, though conditions after birth may vary the details quite a lot."[96] Drawing on Bishop Joseph Butler's account of human morality, she argues that our "nature" consists in a *whole*—a whole constituted by desires, preferences, and affections, along with reason, reflection, and conscience. Our capacities for reason and reflection are not separate from our "nature," they do not impose an alien order on our passions and instincts but are rather one part of our nature and continuous with the rest.

Midgley builds upon Butler's claim that at our "center" we possess a reflective faculty (what Butler called "conscience") that attempts to integrate our passions, desires, preferences, and affections. Conscience decides which of our many desires are desirable. This center, Butler believed, is a natural seat of authority—"our nature itself, becoming aware of its own underlying pattern."[97] Conscience, in its capacity to reflect upon all these conflicting desires, discovers a moral law, a law of our own nature, imposed from within rather than from without. Yet Butler refuses to identify conscience with either reason or emotion. He combines the "moral authority of conscience" with "a proper attention to human feeling."[98]

Our whole nature wishes for integration between the competing claims made upon conscience, and yet freedom from conflict is never attained. Butler's account of morality accords well with what we know about our biological heritage—"His remarks make perfectly good sense in a post-Darwinian context," Midgley argues. "Motivation is fundamentally plural," she contends. It must be so "because, in evolution, all sorts of contingencies and needs arise, calling for all sorts of different responses."[99] Our moral makeup reflects our biological makeup, but the latter does not determine the former. We must negotiate between conflicting values—our bonds and attachments to other living things, our own self-interest and desire—and choose our course of action, but the controller and the controlled in us are part of the same whole.

This model of moral deliberation has bearing on the development of an ethic toward love of other beings, as it does for all our desires and affections. We are inclined to form strong bonds with other creatures but have no closed instincts to act as an infallible guide to action. Our instinctive tendencies can be good or bad, depending on their context and scope. Midgley does not discuss Butler's view of love, but his arguments here are instructive. Butler believes that we have strong social affections, yet doubts that it is possible for humans to love everyone in our community (even if our community includes only other humans). Compassion, however, is much more widespread, owing to its particular social function—it leads humans to help one another and conscience recognizes its value. Butler's arguments here take the form of "functional arguments," such as those used in biophilia. Just as our eyes are adapted to seeing, so is "the nature of man adapted to a course of action." But love and compassion for others can also be overindulged, misguided, and disproportionate, as any other of our passions can be. Our social affections have their "stint and bound," as do all affections, and we must reflect carefully upon them.[100]

Like Wilson's, Butler's theorizing about human nature proceeds by inquiring into the function of the various features of our moral constitution, both "good" and "bad," but the picture that emerges is one of greater moral complexity. Such complexity makes perfect sense in light of our biological heritage. Our evolutionary past created ever wider possibilities for behavior and, with these, more rather than less conflict between our divergent inclinations. "We want incompatible things," Midgley observes, "and want them badly."[101] The best we can do is to reflect at length upon all these values and arrive at some means of ranking them. Human intelligence has evolved in response to the conflict in the world around us and the world within us, producing our own incompatible desires. Animals too experience these conflicts of interest, but humans appear to have evolved a greater ability to understand "what is going on," Midgley argues, and to use this power of understanding "to regulate it."[102]

To sum up, we have moved from a discussion of biological accounts of the human bond with animals to an argument about the moral complexity inherent in human ethics. What connects these two topics is the idea of nature in its broadest sense—the natural world around us and our own human nature. Earlier, in comparing Darwin's understanding of nature to the perspective of land ethics and theocentric ethics, my concern was to illustrate the inevitability of conflict and strife in nature. In discussing human "nature," my intention is essentially the same, namely, to suggest that our moral nature, too, is characterized by conflicts that must be adjudicated but can never be resolved once and for all. The connection between the two ideas, as I see it, is as follows. In our responses to natural environments our effort to make decisions that will produce "good" outcomes is, or ought to be, constrained by natural processes themselves *and* by a recognition that, as humans beings, our moral reasoning is inherently fraught with ambiguity. We are torn between values that cannot be harmonized. We are equipped with responses that may be described as love, for example, but we cannot always or easily rely upon these for direct action. Deliberation about what constitutes an appropriate set of obligations must take place, and these obligations may or may not match up with the ends we desire. Land ethics undertakes this sort of deliberation with regard to specific contexts; so does theocentric ethics. Let us consider, then, how a response of love may translate into moral action in environmental ethics.

A QUALIFIED, NATURALIZED ETHIC OF LOVE

If there is one thing that Midgley, Gustafson, Wilson, and Rolston all agree on, it is that we often experience a strong pull toward animals and nature. But these authors are much more emphatic about the need for deliberation on how love translates into an appropriate course of moral action. "Human valuations," Gustafson writes, "grow out of our natures; there are 'natural' and cultural bases for the desires we have and the ends that we choose." But, he adds, reference to nature "does not imply that conformity to nature in a simplistic sense determines the proper objects of love or the proper intensities of valuations."[103] Adopting a comprehensive naturalized approach implies that we can extend an ethic of love to nature, but it also implies that we need to supplement it with other considerations about appropriateness—love's stint and bound, as Butler says. Our natural propensities (social bonding, neoteny, biophilia, if it exists) come into play with both wild and nonwild nature, but in the case of wild nature we may have to *refrain* from expressing love in a direct way. We have to view our own internal "nature"—our natural tendencies to love other animals—within the context of external nature—the natural processes that are to be respected. In other words, if we have a natural love of wild nature, it must be qualified by considerations of what is natural *for nature* as well, and the two may not coincide.

We may want to intervene in many cases where consenting to natural processes suggest that we should not, as with wild animals whose plight is part of natural conditions. With domestic animals we are not so restrained. Here it seems our tendencies to bond can often be freely expressed. This freedom to love and be loved unconditionally—a *virtuous* "excess" of a loving expression, as Stephen Webb has termed it—may well be the "hallmark" of human relationships with animals we keep as pets.[104] In what follows I will suggest some of the ways in which an appropriate ethic of love might issue guidelines (both in terms of action and refraining from acting) for humans' encounters with wild and nonwild nature as well as the places where human and natural communities intersect.

THE PARAMETERS FOR AN ETHIC OF LOVE

In thinking about how we can love nature we first need to discriminate between different types of living things. Evolutionary continuity may imply moral consideration for all things, but it does not imply the same *kind* of treatment for all living things. As Midgley argues, in order to know how to treat another being, it is crucial that we know what kind of organism it is. Differences can be real and "need to be respected for the dignity and interests of those most closely involved in them."[105] Discrimination along biological lines is not necessarily an inappropriate or prejudicial discrimination. Of course, some ecotheologians do discriminate between different kinds of organisms: Birch and Cobb, for instance, make distinctions between those with and without richness of experience. But this is discrimination along the *wrong* lines, without consideration for the natural value of living things or their specific needs as wild or domestic animals.

McFague too gestures toward respect for the distinctiveness of different living things (despite the fact that all are to be treated as subjects analogous to human subjects.) She argues that in *re*training the objectifying eye to see others as subjects, "part of our education is to focus on the distance, the difference, the particularity, the uniqueness, the 'in itselfness,' the indifference, the otherness of the other."[106] This reminder that each animal is valuable for its otherness sounds promising, yet her overarching concern with the suffering and oppression of animals minimizes differences between them that need to be respected. In our dealings with some kinds of organisms, as land ethics correctly asserts, natural suffering must be left to take its course. In any case, even if we grant that McFague's ethic can honor differences between all living things, it is difficult to get any workable ethic from an exhortation to appreciate the uniqueness of *each individual* organism. She offers no guidelines for recognizing what these differences are and how they might entail distinct ethical obligations. We need general guidelines about different types of beings—species or wild/domestic classifications. Loving each creature for its uniqueness does not constitute an ethic.

Some of the most important distinction to be made, then, are those between wild animals and domesticated ones and between native species and introduced or exotic ones. Native species ought to be preserved not only out of a sense of gratitude for the given ordering but also because of the often devastating effects of introduction. While it is difficult to demonstrate with precision the impact of such introductions, circumstantial evidence of environmental devastation is "overwhelming": "Many extinctions, especially of birds and plants but also mammals, reptiles, snails and others have been attributed to introduced species."[107] In the language of land ethics naturalness is to be valued; in the language of theocentric ethics we consent to natural processes we did not create and do not control. Compassion as an ethic toward wild animals is not discriminating enough so long as it fails to make these kinds of distinctions. It is "species blind in a bad sense, blind to the real differences" that exist between different types of organisms.[108]

LOVE FOR WILD AND "SEMI-WILD" NATURE

An ethic toward wild nature entails what Rolston calls a love of life beyond self-love, a love of something external to and in many ways unlike ourselves. Love for wild nature is not simply a variety of self-love because it is not founded on sameness with humans. Rolston's account of certain "charismatic" species, those possessing giftedness—a skill and virtuosity distinct from our own—is one expression of a love of wildness. As we have seen, the actions that flow from love for other species must still be worked out with regard to other values and the specific context in which moral action is required. Midgley too realizes that the ethical issue of how we are to treat animals raises an important "philosophical point about the otherness of others. Can we be concerned at all with anything which is not fundamentally the same as ourselves?" Her answer is yes: "What would a world be like in which we only cared for others in proportion as they were like ourselves?" What is it like, she asks, to be an incubating gull or an emperor penguin?[109] Midgley's question is somewhat rhetorical. She does not go on to answer it as ecotheologians often do by locating similarities between ourselves and other living things.

Perhaps most significant, the idea of charismatic species embodies the idea that life-forms are not only *gifted* but also, in a sense, *gifts;*[110] that is, the term *charismatic* points to their value and origin as something external to ourselves—*kharisma* meaning that which is divinely given. In this sense, all species, and indeed all of nature, is charismatic. Love of wildness and a response of gratitude ought to be inseparable. We can think of wild beings as neighbors to be loved, if we keep in mind that their nature and value exists outside of human ordering but "not outside both divine and biological order."[111] Love of wildness acknowledges there are "vast ranges of creation that now have nothing to do with satisfying our personal desires" as well as the vast stretches of evolutionary time that preceded our arrival on earth. What wild nature teaches us is that "God is not for us alone."[112] Our understanding and appreciation of natural processes point to the conclusion that we are not the center of things. This discovery—supported by biological evidence—is itself an important and profound gift, ethically and theologically.

Is it the case that love for wild nature never finds individual expression—that we cannot feel love for a particular wild being in a particular situation? Not necessarily. On the one hand, love of wildness for what it is implies valuing other beings without seeking to discover similarities to oneself in the object of love, on the other, it also means valuing the other without being able to possess the other. We saw that Leopold criticized the "primitive" mentality of the person who appreciates nature only for its value as something to be possessed, invaded, and appropriated. Love of wildness relinquishes these claims, without relinquishing moral (or aesthetic) responses to wildness. This, it seems to me, is also how Terborgh regarded the injured peccary. He kept his distance from it, at least partly out of respect for its naturalness. Here we find a case of love toward an *individual* animal that is nevertheless constrained by considerations of what is natural for the individual and what is best for the species as a whole.

As I have argued, love and gratitude are starting points for determining our ethical obligations; they provide a motivational basis for moral action as well as moral restraint. In keeping with land ethics and a theocentric orientation, love recommends participating cautiously in natural processes. It also recognizes, however, that it is often impossible for us to preserve the wildness that is loved by doing nothing, because our previous actions have already compromised natural values. Therefore, loving wild nature is not simply letting it be. A general response of love requires specific actions. In wild nature native species (animals and plants alike) are given priority over introduced ones, and endangered species deserve special consideration and care. The species as a whole is generally more important than individuals. But species must also be understood within their larger ecological and evolutionary contexts. A response of gratitude is owed to the natural processes themselves as much as to their products.

A disposition of love and gratitude for the given ordering is directly relevant to real cases in wildlife management practices. We have seen that paying inordinate and inappropriate attention to preserving individual organisms, rather than species or ecosystems, disrupts the very patterns of interdependence that environmentalists value. Focusing on the species unit broadens the ethical scope, but even this focus is not sufficiently broad if it does not include consideration of the larger context in which species exist. For example, when a species is threatened or endangered because of habitat loss, appropriate wildlife management endeavors to preserve the habitat not only of the threatened species but of that upon which it preys as well. In protecting the threatened northern goshawk, the Forest Service also takes into account the preservation of the habitats of robins, band-tailed pigeons, woodpeckers, squirrels, and other goshawk prey species.[113] Managers are not in a position to issue judgments about which of these bird species is the most valuable in terms of mental or physical capacities, or to ask whether it is a "problem" that the existence of the goshawk causes suffering and death to other sentient organisms. Dependence and conflict are inseparable, and a love of wildness consents to this arrangement. Often the situation in wildlife management is far more complicated than this, however, and its complexity is exacerbated by past interventions. Not only must consideration be given to both predator and prey in a given habitat, but it may turn out that the protection of one threatened species has adverse effects on another threatened species. Preservation of the northern goshawk, for example, has resulted in losses to the threatened Mexican spotted owl, both because the goshawk requires different forest conditions and because it preys upon the owl. Cases of extreme conflict between conservation strategies for two endangered species are often the result of past management practices that have focused too much attention on one species at the expense of others. These "individual management plans" have created situations of great moral complexity; attempts to resolve these conflicts "will have to be made on a case by case basis."[114] Here, our past actions have had unintended consequences and have curtailed our present ability to do what is "best" for all species involved. Commenting on the guidelines spelled out in the Endangered Species Act, the National Research Council acknowledges both the possibilities and the limitations that shape our interventions into nature: "The past provides opportunities for the future but also constrains it" (16). This perspective is consistent with a theocentric interpretation of human moral action and the conditions of responsibility as well as finitude that guide it.

Conflicts, whether anthropogenic or natural, cannot be completely eliminated, and a disposition of love and gratitude accepts this. Generally, protecting large areas of habitat offers the best protection for species that are now threatened by human activity, and this action is recommended for the goshawk-owl conflict. Indeed, preservation of one species may have much wider ramifications, "inadvertently affect[ing] dozens of nontarget species found in the same habitats" (116). Greater knowledge regarding the workings of ecological systems and the habitats of various species increases, rather than decreases, the likelihood that environmentalists will have to deal with such cases: "The potential for such conflicts will rise as ecologies of listed species become better known" (120). Still, close studies of natural processes, combined with an awareness of our own limitations in doing what is best for nature, remain the tools available in wildlife management. Strategies thus involve "multispecies plans" and "habitat mosaics" with an eye to preserving this larger whole. But such plans cannot always respond to declines in every species in an ecosystem, and difficult choices have to be made. In such cases preservation efforts are sometimes guided by a triage approach, based upon ecological considerations such as whether the "loss of one species [would] have a greater effect on the ecosystem than the loss of the other" (121). As noted in chapter 2, triage sometimes abandons the most desperately imperiled, and as such it is at odds with an ethic of love such as McFague's that gives preference to the neediest in nature.

As these examples suggest, a love of wildness entails an acceptance of the fact that tragedy and conflict are an inevitable part both of nature itself *and* of human moral action. In this sense "management" of wild lands does not undermine the stance of participation, even though the former might seem to suggest a more controlling activity than the latter. At the heart of wildlife management lies a recognition that our current interventions are our best (but imperfect) effort to compensate for past actions. Management is not simply a resuscitation of old Cartesian attitudes of control that ecotheologians (and at times Leopold as well) have decried—because it is not an end in itself but a means of allowing natural processes to continue. Wildlife managers "manage in order to compensate for human interruptions."[115] Moreover, at no time does such management assume that humans and their concerns and values are at the center of decisions and strategies; management and participation both attempt to discern what actions are appropriate, given a vast multidimensionality of values.

It might be argued that managing wild nature is a contradiction in terms; where management is occurring, nature cannot be considered "wild" in the sense of remaining unaffected by human activity. Rolston in fact gestures toward this paradoxical and contradictory nature of wildlife management in some of his comments. But, as Robin Attfield argues, nature that is wild in this sense would not *need* to be managed. There *is* something paradoxical, he notes, "concerning human agency in restoring what is essentially independent of human agency. But it should also be noted that there is no question of this where nature is undamaged and wilderness untrammelled."[116] The paradox, in other words, is not a logical contradiction. Management interventions in (truly) wild situations might be motivated by the need to *prevent* activities that would damage wildlands, but, Attfield notes, this is not the same thing as rehabilitating or restoring wild places.

There is, however, something tricky about the idea of restoring wildness, for the concept raises the question of whether a wild place, once altered, can ever be made "wild" again. We have seen that Leopold considered wilderness something that could shrink but not grow. Does this mean that wildness, once compromised, is lost forever? Not necessarily. Leopold distinguished what he called "artificialized" management (such as the extermination of predators) from land ethical management, which seeks to work with natural processes. Thus while loss of value occurs when wilderness is destroyed, those values can, to some extent, be replaced. This is not to argue that humans themselves create natural value but, rather, that human activities can at least provide conditions under which nature creates, or re-creates, those values. Attfield argues that the paradoxical element involved in such collaborations between humans and nature in re-creating value is "perhaps no stronger than that involved when the agency and interventions of parents bring about the possibility of autonomy for their children as they grow to adulthood; though here autonomy is of course initiated rather than restored."[117]

Perhaps a more crucial question for an applied environmental ethic inspired by love for wildness is whether or not the average person has any role to play in participating in natural values and processes.

I suggested at the outset that knowledge of science and environmental ethics should not be the exclusive province of professional scientists, however much they may resent the intrusions of well-meaning laypersons. While actual decisions regarding wildlife and wildlands should, and, I think, must, be made by professionals in the field, there is still much to be done by people who are not scientists and who live in areas where human culture intersects "nature" in some form. What sort of guidelines might a qualified ethic of love for nature provide for the vast majority of us who are *not* foresters or park managers? One very promising movement that seeks to join human communities and natural "communities," while respecting natural processes, is bioregionalism.

Bioregionalism is a movement that emerged in the 1970s from philosophers and ecologists such as Peter Berg and Raymond Dasmann and, in recent years, has been endorsed by environmental writers such as Gary Snyder, Wendell Berry, and Scott Russell Sanders.[118] Bioregions are defined as distinct geographical areas, with particular types of landscapes, soil composition, native species, and climatic features, but they are also places where nature and culture intersect, where people make their *homes*, in the deepest sense of the word. Being at home is as much a spiritual and psychological rootedness as it is a physical or geographical one; bioregionalists encourage people to "reinhabit" the places in which they (already) live by coming to know, understand, and appreciate what is unique about those places and their history.

Bioregions link cultural, regional, and ecosystemic interactions, consistent with Leopold's conception of land ethics as enlarging "the boundaries of community to include soils, waters, plants, and animals, or collectively: the land."[119] Proponents of bioregionalism encourage us to learn more about the natural history of places in which we live in order to understand how best to live there—that is, how to live there without compromising the native species or the fundamental character (what René Dubos calls the "genius") of the place. Miles of green golf courses in naturally dry, desert states clearly do not respect the genius of a place, nor are such developments sustainable. Bioregionalism also critiques the homogenization of towns and cities resulting from the spread of fast food chains and superstores—the cultural/economic equivalent of invasive species that destroy native habitats and alter the innate character of landscapes.[120] But bioregionalism has a positive agenda as well: it promotes a form of practical land ethics that reinforces a feeling of love for and identification with a particular place—a sense of "living-in-place"—and provides a set of guidelines that include restoring native plants and animals to an area, buying locally grown or produced products, learning more about natural resources, waste recycling, weather patterns, and watersheds in one's area. Bioregionalists practice Leopold's rule of saving all the parts of a biotic system, but restoration, not just conservation, is part of the objective. As an environmental movement it encourages scientific familiarity and study of places as well as ethical responsibility in terms of cultivating the civic virtues necessary for becoming what Leopold called true citizens, rather than conquerors, of the land. "In a very real sense, bioregionalism seeks to fulfill Aldo Leopold's notion of a 'land ethic.'"[121]

The conceptual and philosophical details of bioregionalism are still developing, but many proponents see affinities between this orientation and a religious sensibility. Certainly, Gustafson's and Leopold's emphasis on humility, gratitude, and respect for nature is compatible with a bioregional approach that attempts to evoke these virtues as responses to *particular* places. Some environmental scholars have identified Leopold (as well as Rachel Carson and Henry David Thoreau) as advocates of an environmental virtue ethic that links the development of human character, flourishing, and practical wisdom to moral action within, and on behalf of, specific biotic communities.[122] Whether or not such virtues are consistent with the response of natural piety, or the sense of wonder and divinity, that inhere in theocentric ethics and land ethics is a matter of personal perspective; bioregionalism may be either a secular or religious stance. As it continues to develop, its practitioners need to be wary of defining bioregional objectives in overly simplistic terms, such as being in "harmony" or "balance" with nature (or conversely, defining environmental vice in terms of disrupting naturally harmonious nature). Proponents of bioregionalism must also be careful not to elide the differences between human communities and natural communities, even while encouraging appropriate forms of participation in natural processes. The definition of bioregions must also take seriously developments in evolutionary ecology that highlight the difficulties involved in defining biotic systems with precision, whether in terms of geography or history (i.e., ecosystems are neither clearly bounded nor completely stable over time). But, on the whole, this approach has great potential: it captures both the dispositional and the practical dimensions of Leopold's sense of love for the land, and it retains his presumption in favor of respecting natural processes. This approach also coheres with some of the imperatives of ecotheologians (particularly, I think, McFague's arguments) that we work harder at seeing the world around us with different eyes, that concepts of knowing and loving become more closely aligned. But it has the added, practical advantage of allowing discriminations such as those between native and nonnative species and natural and unnatural elements of a region.

LOVE FOR NONWILD NATURE

Though sharp distinctions cannot always be drawn between what is wild and what is not, love of wildness has a different agenda than love toward domesticated animals and pets. Here compassion is appropriate and warranted. Whereas the former suggests that we sometimes let nature take its course even when this involves suffering, nonwild animals make claims upon us that *are* founded in some sense upon their "sameness" with us, although their similarity has less to do with genetic or physical similarities than it does with similarity of *context*: domestic animals are a part of human culture. Wild animals are members of a *biotic* community; domestic animals belong to what Midgley and Callicott refer to as the "mixed community," falling within a "spectrum of graded moral standing" that includes "family members, neighbors, fellow citizens, fellow human beings, pets, and other domestic animals."[123] At least some of the ethical imperatives issued by ecotheologians (loving, healing, feeding) are appropriate, if they are confined to this sphere of beings and not extended to nature and animals as a whole. Basic responses and impulses to love nonwild creatures can be given freer rein.

Whereas land ethics, in keeping with its evolutionary and ecological underpinnings, does not aim at ensuring the well-being of each organism in the biotic community, an ethic toward this mixed community does care for individual members. We protect our domestic animals and pets from wild predators even while we do not stand between predator and prey in wild nature. "Where animals suffer owing to human domestication, they are removed from nature, and compassion is warranted."[124] Humans as a whole have entered into a relationship of trust with animals that have been domesticated. This relationship, even when it involves instrumentalization of animals, is not inherently oppressive and probably has ancient, evolutionary origins, as I have previously suggested. Animals and humans have drawn one another into a relationship that can be mutually beneficial.

Even though the process by which animals have become domesticated may have "natural" origins in our evolutionary past, domesticated animals are now more the products of culture than of nature. To the extent that we have taken away their wildness, we owe domestic animals protection, compassion, and—especially if they are our pets—love. Again, our *prior* participation and intervention in nature, which has put these animals in a state of vulnerability and dependence, created obligations we are now bound to uphold. To turn domestic animals loose, at this point, and allow them once again to become wild (or more accurately, feral) would probably only create more suffering and death.[125] Given the relationship we have created with them, our treatment of all animal members of the mixed community resembles our treatment of other humans: we treat them humanely, in the literal sense of the word.[126]

To include domestic animals as members of human communities is not to deny them any individuality or otherness, nor does it imply that they possess no remnant of their once wild nature. Even when we treat animals humanely and include them as members of our culture or community, we nevertheless continue to treat them in a manner that takes their natural characteristics into account. We should respect as much as possible the specific natural preferences of even domestic animals—their need to run, to interact with members of their own species, their dietary requirements, and so on. Individual preferences can be respected, but on the whole we regard domestic animals with the same basic principle of compassion that we extend to other humans.

A general policy of valuing nature's (and animals') "otherness" can be as inadequate as valuing beings according to their sameness to us. Ethical obligations stemming from our moral inclinations demand a more critical reflection than they often receive in environmental literature. It is not enough that we respond to the otherness, the complexity, the richness, or the suffering of other beings; we have to ask whether our response is appropriate. Answering this question involves a process of moral deliberation and comparison of values that few ecotheologians, or secular environmentalists, have fully appreciated.

If ecotheologians took their own insistence on interdependence (it all its senses) more seriously, perhaps they would see that the ethics they promote do not follow from this concept. Following nature in an uncritical fashion distorts our understanding of the way in which natural processes operate and the appropriate role for humans vis-à-vis our natural environment. The work of Midgley, land ethicists, and Gustafson points to the need for a more comprehensive naturalized ethic. Inferences from the natural world, such as the idea of interdependence, are important, but, in and of themselves, they clearly do not "furnish a test to resolve our clashes of value."[127] On this science and theology can agree. I have attempted here to offer the outlines of an alternative approach to environmental issues, one that draws on elements of Darwinian natural science as well as theological orientations; I have argued that appreciation and gratitude for a broader set of values (i.e., a nonanthropocentric understanding of value) implies a more finite range of moral action in our dealings with the natural world. With this perspective in mind, I have suggested some of the ways in which an approach of love toward nature may still be appropriate, so long as it is qualified with reference to that broader spectrum of values. Thus the approach to environmental problems presented here endeavors to make sense of our intuitive moral responses to nature and the relationship between these responses and appropriate ethical action in the natural world.

CONCLUSION

Finitude and Responsibility

In the last few decades the rise of ecological theology and the discipline of environmental ethics as a whole has signaled an important effort to shift concern toward nature and nonhuman forms of life. Ecotheology stems from the conviction of many Christian thinkers that the beliefs, practices, and paradigms of their tradition can and ought to be developed along environmental lines. The imperative to do so is strengthened by scientific evidence pointing to a fundamental *interdependence* of all life. One of the primary tasks that I have undertaken throughout this project is to closely examine the use of ecological and evolutionary concepts in environmental ethics in order to see whether they cohere with current scientific—particularly Darwinian—perspectives. As I have argued, many environmentalists have not incorporated accurate scientific knowledge into their arguments, despite their claims to the contrary.

The concept of interdependence plays a crucial role in much of environmental ethics and has therefore served as a point of departure for many of the arguments in this study. I have attempted to distinguish the different meanings and usages of such concepts in environmental arguments and to subject these concepts to both scientific and theological scrutiny. In so doing, several related topics have been addressed, among them the proper relationship between nature and ethics, the appropriateness of particular guidelines (loving, healing, liberating nature) that these allegedly "naturalized" ethics generate, and the problematic persistence of anthropocentrism in ethical norms and values regarding nonhuman life.

The ecological model that pervades much of ecotheology is assumed to embody both a religious and a scientific interpretation of the value of nature and the place of humans within it. Nature is often understood as both the source and the object of ethical norms; the ecological ethic, as we have seen, is both derived from and extended to nature. Nature presents us with an ideal community, according to some of these arguments. At the same time, we are told, Christians have an obligation to establish and enforce harmonious community in the natural world, and the world as a whole. These arguments, I have contended, misunderstand nature (in its broadest sense) as well as aspects of *human* nature (in terms of the appropriateness and the possibility of our moral governance of the natural world).

As I have argued throughout, misinterpretation of "nature" (in both senses) is intimately connected to a neglect and/or misuse of basic scientific data. Much of environmental ethics can be criticized from a scientific perspective, as I have demonstrated, but theological perspectives reveal many of the same inadequacies. From the perspective of science, ecotheologians, as well as some secular ethicists, borrow selectively from scientific data in order to bolster particular ethical claims. One example of this is seen in the emphasis upon evolutionary continuity as a key insight in much of environmental ethics, secular and religious alike. The fact of evolutionary continuity of all life-forms is frequently highlighted. Nonhuman forms of life are to be respected according to their possession of certain (evolutionarily produced) traits, including consciousness, sentience, or complexity, even while the forces that produce these traits are ignored or deemed evil in some sense.

Philosophers such as Singer and Regan attempt to break out of what they see as a "speciesist" bias against animals, arguing that animals instead be understood as "persons" or "subjects" akin in certain respects to humans. Ecotheologians such as McFague, Birch, and Cobb also draw our attention to the presence of humanlike subjecthood in nonhuman life: we are to understand these beings in terms of a "subject-subjects" relationship; animals make moral claims upon us in proportion to their capacities for rich experience and intelligence. Thus the basis for moral concern is often expressed in terms of anthropocentric arguments and categories that emphasize the similarities, both physical and mental/psychological, of humans and other life-forms. This use of evolutionary data is problematic from a scientific standpoint in that it overlooks significant *differences* between humans and other animals (i.e., the biological significance of species), fails to separate moral claims of wild and nonwild animals, and pays insufficient attention to the natural (or non-natural) contexts in which various life-forms exist.

Ironically, then, the idea of evolutionary interrelatedness is often promoted without regard to interdependence in terms of *interconnectedness*—the evolutionary processes, the struggle and strife, and the trophic and ecosystemic interactions that are an integral part of nature. Like evolutionary relatedness, interdependence as interconnectedness finds uncritical support in much of environmental ethics. Nature's interdependence provides a model for community life, an allegedly scientific rationale for loving, liberating, and healing the suffering of nature and animals.

In ecotheology interdependence is often used interchangeably with such terms as *community*, *mutuality*, and even *harmony*. This is seen most clearly in Ruether's account of nature and her call for a conscious emulation of its nonhierachical, symbiotic structure. Moltmann, too, praises the harmonious and provident functioning of ecological communities (even while hoping for the liberation of all life from conditions of suffering). In general, the ecological, community ethic takes the health and well-being of the individual—human or nonhuman—seriously, while recognizing the importance of the *whole* as well. Some secular ethicists, on the other hand, deny the possibility that a larger whole can be the proper object of moral concern or a repository of "rights" or "interests." Emphasis is instead placed on locating value in, and directing moral concern to, the lives and suffering of individual sentient beings. However, among ecotheologians it is often assumed to be desirable, and possible, to promote the good of the ecological community while attending to the needs of individual beings. Longing for a perfect and permanent harmony in nature, ecotheologians invoke the ecological model in yet another sense—namely, in terms of a theological, eschatological, model for nature as it once was—and will one day be again.

As I have argued, there is a great deal of ambivalence regarding the *true* nature of nature and considerable confusion about whether or not nature already exemplifies the ecological model.

From a scientific standpoint this interpretation of an interdependent, ecological community cannot be sustained: nature does not provide for individual beings; interdependence in nature is itself the source of much conflict and struggle, *not* an overriding, harmonizing principle as theologians such as Moltmann suggest. Ecotheologians' interpretation of interdependence fails to recognize that the good of the parts and the good of the whole cannot be harmonized.

On the whole, interdependence (whether understood as interconnectedness or interrelatedness) does not provide us with a clear or unambiguous norm. For this we have to turn *inward*—to an understanding of human moral reasoning and moral complexity—and *outward*—to a source of values beyond strictly anthropocentric values. Both of these turns, I have argued, present us with more conflict and ambiguity than ecotheologians and secular ethicists have realized. Gustafson's theocentric ethics, land ethics, and Midgley's arguments provide a more accurate account of the conflicts inhering in internal and external nature. Thus, as I have argued, ecotheology can be criticized on theological and philosophical grounds as well as scientific grounds.

From a theocentric and ecocentric perspective the failure to come to terms with nature as it *really is* constitutes a fundamental lack of respect for the natural ordering and processes that sustain, as well as destroy, life. As Gustafson argues, the belief that we can control or alter the given ordering implies a denial of God as God, a refusal to consent to the powers that manifest the Deity in the sphere of nature. From the perspective of land ethics we are similarly urged to respect evolutionary processes (not just their products) and intervene only as necessary to allow biotic systems to function naturally. Yet neither Gustafson's ethics nor land ethics amounts to a simple imperative of "following nature" or "letting nature be." Rather, given the complexity of our responsibilities (including our prior interventions in nature), as well as our *own* nature as prioritizers of value, environmental ethics necessarily engages us in a difficult and complicated endeavor. I have argued for a general presumption in favor of respecting natural processes and intervening cautiously, with regard to specific details, the conflicting values at stake, and the broader ecological context in which values are embedded.

Along with theocentric ethics and land ethics my arguments affirm a notion of humans as participants in nature. We are finite beings, subject to certain conditions in the external world and limited by our inability to perceive the larger good that we may seek to preserve. In part the reasons for viewing humans as participants, rather than controllers, lie in nature itself: our dependence upon natural patterns and processes is a fact, regardless whether this fact is construed religiously or otherwise. But other grounds for understanding our role as participants can be located in arguments about human nature. As Gustafson argues, moral action is contingent upon a ranking of values—some interpretation of the relationship of parts to wholes. Yet, as I have already suggested, we cannot readily apprehend "the whole" or discern with certainty what is "good" for it. This interpretation of our own finitude suggests that we assume a more modest role in our attempts to oversee the natural world. The concept of humans as participants, as I understand it, embraces moral modesty.

Given this account of nature and human nature, and of theological, philosophical, and scientific perspectives, I have attempted to articulate a more appropriate environmental ethic, demonstrating how a response of love may be translated into ethical action, with regard to a number of important distinctions. Human moral (innate, intuitive) responses to nature have merit, but such responses must be tempered and qualified by further deliberation. Responses such as love and concern for suffering (whether driven by our biological makeup or theological convictions) need to be critically examined before they are extended to other life-forms. First and foremost, consideration must be given to the *kinds* of beings with which we are dealing. An indiscriminate love ethic is not a virtue when applied to nature. Love toward nonwild beings takes a different form and expression from love for wild nature. A compassionate response to suffering is appropriate in some cases (animals in cultural contexts) but not necessarily in others (e.g., nonendangered, wild animals whose suffering results from natural causes). Sources of suffering—natural and anthropogenic—must be clarified. Moral concern cannot simply fall along lines of sentience or degrees of complexity, without regard to diversity of life. In general, species themselves, not individual animal lives, must be valued and preserved in natural environments.

An appropriate environmental ethic turns on distinctions. In attempting to do what is good for nature, we first must ask, "Good for whom?" "Good for what?" The answers to these questions, and the moral actions that follow from our answers, will necessarily reflect the complexity and ambiguity of the world we inhabit as well as our own status as finite, yet responsible, moral agents.

NOTES

INTRODUCTION

1. Darwin's account of moral evolution influenced Aldo Leopold's understanding of our development of an "ecological conscience" that is crucial for land ethics. The theory of natural selection also plays a central role in land ethics as a directive for our treatment of wild nature. I will return to both these points.

2. Darwin, *The Origin of Species*, 450.

3. For his own part, Darwin suggested in the *Descent of Man* that feelings of sympathy for other creatures were themselves an evolutionary response, thus strengthening the connection between evolution and concern for animal welfare.

4. At a House Judiciary Committee hearing on May 10, 2000, proponents of "intelligent design theory"—which holds that there are irreducibly complex natural systems that Darwinism cannot explain—catalogued the negative effects of Darwinism on education, religion, and morality. During their testimony some speakers invoked a recent work, Thornhill and Palmer's *A Natural History of Rape* (which explains the origins of rape in terms of adaptive reproductive strategies), as an example of the link between Darwinian theory and moral decline. Defenders of intelligent design also characterized Darwinism as a dogma of materialism, closely guarded by a scientific priesthood that will brook no criticism of the orthodox view.

5. I suspect, however, that these broader concerns about the social and moral implications of Darwinism may contribute to the neglect of evolutionary theory in much of environmental ethics. In particular, the interpretation of Darwinism as "survival of the fittest" seems to be at odds with the concern among some ecotheologians with the oppressed and marginalized (both human and animal). It is difficult to reconcile an ethic of justice and liberation for the dominated with a reading of Darwinism that seems to legitimize domination of the "weak" by the "strong." I will return to this point briefly in the next two chapters.

6. Certainly, there is more "lay" interest in ecology than most other sciences—and perhaps an unwarranted assumption among its lay devotees that ecology is accessible to all. Interestingly, there is generally no self-censure among scientists, however, many of whom often assume that religion, ethics, and theology are within their grasp, despite having no formal training in these disciplines.

7. While scientists acknowledge there to be a "background extinction rate" that occurs naturally, humans have increased that rate, "perhaps by orders of magnitude." National Research Council, *Science and the Endangered Species Act*, 5.

8. In a recent essay entitled "Ecological Confusion Among the Clergy" Allen Fitzsimmons notes that this concept of ecological sin, and other nonscientific pronouncements regarding nature by clergy and religious organizations reveal a level of "ecological misunderstanding and theological confusion" that could lead to "misguided policy." Fitzsimmons, "Ecological Confusion Among the Clergy," 1.

9. McFague, *Super, Natural Christians*, 164.

10. Ibid., 41.

1. THIS VIEW OF LIFE: THE SIGNIFICANCE OF EVOLUTIONARY THEORY FOR ENVIRONMENTAL ETHICS

1. Gould, *Ever Since Darwin*, 12. This is a surprising statement coming from Gould who, as we will see, maintains a strict separation between scientific facts and religious or ethical values (for example, see his more recent work, *Rocks of Ages*.)

2. Intelligent design theorists obviously disagree.

3. Gould, *The Panda's Thumb*, 27.

4. White, "The Historical Roots of Our Ecologic Crisis," 1203–1207.

5. Although some environmental ethicists use the terms *ecocentric* and *biocentric* interchangeably, I give preference to the former term throughout this work. A biocentric, or "life-centered," approach may take into account characteristics of nonhuman life (and locate values in those characteristics) yet fail to understand these values in a holistic or systemic fashion. In *Respect for Nature* Paul Taylor, for example, understands organisms as "teleological centers" with inherent value of their own, but his ethic remains focused on individual lives and is more biocentric than ecocentric.

6. For instance, see Sagoff's essay, "Animal Liberation and Environmental Ethics."

7. Whether or not treatment of domestic and lab animals is appropriately considered to be an *environmental* concern, it is certainly the case that our treatment of these animals can have huge consequences for the environment, as when waste from factory farms contaminates water supplies or when genetically modified animals escape and interbreed with wild members of their species, jeopardizing the survival of the species as a whole.

8. For a summary of Gould's position on this and other debated issues in contemporary evolutionary biology, see Sterelny, *Dawkins and Gould*.

9. Ibid., 76.

10. Gould's critique of conventional Darwinian theory is spelled out in detail in his magnum opus, *The Structure of Evolutionary Theory*.

11. Dawkins has addressed some of these criticisms in works subsequent to *The Selfish Gene*. Still, his choice of title is an unfortunate one if he wishes to avoid accusations of genetic determinism and anthropomorphic interpretations of genes.

12. Ibid., 54.

13. The Gould-Lewontin critique appears in "The Spandrels of San Marco." Gould has discussed "spandrels" (nonadaptive structures in evolution) elsewhere, including "Fulfilling the Spandrels of World and Mind."

14. Wilson, *Sociobiology*. For an account of the sociobiology debate and the characters involved, see Segerstrale, *Defenders of the Truth*. Wilson offers his own version of these events in his memoirs, *Naturalist*.

15. Thornhill and Palmer's *A Natural History of Rape* is one example. For a more recent, and very different, take on evolutionary psychology and human evil, see Williams, *Doing Without Adam and Eve*. Williams interprets the Genesis account of the fall in light of

sociobiology and attempts to reconcile theological accounts of sin with biological explanations of human evil.

16. Sterelny, *Dawkins and Gould*, 51–52.

17. A modified and more limited argument for group selection is endorsed currently by David Sloan Wilson.

18. As David Hull notes, reports vary regarding the exact amount of water dumped on Wilson during this incident. Segerstrale recalls a jug; Wilson maintains it was a pitcher. For his own part, Hull remembers only a "small paper cup of water," adding that his version is strengthened by that fact that "Wilson was able to mop up the water with a single handkerchief." Hull, "Activism, Scientists, and Sociobiology," 673–74.

19. Gould, *The Structure of Evolutionary Theory*, 7.

20. Worster, *Nature's Economy*.

21. Ibid., 10.

22. Linnaeus, "The Oeconomy of Nature," cited ibid., p. 36.

23. Bowler, *Evolution*, 60.

24. Worster, *Nature's Economy*, 52.

25. Bowler, *Evolution*, 60.

26. Worster, *Nature's Economy*, 39.

27. Ibid., 82.

28. Ibid., 84.

29. Marshall, *Nature's Web*, 334–348.

30. Golley, *A History of the Ecosystem*, 3.

31. In *The Greening of Protestant Thought* Robert Booth Fowler discusses the prevalence of the community metaphor for nature among Protestant thinkers.

32. McFague, *Super, Natural Christians*, 151.

33. Ibid., 45.

34. Ibid., 38.

35. Hagen, *An Entangled Bank*, 13.

36. Ibid., 21.

37. Golley, *A History of the Ecosystem*, 201. Hagen and Golley disagree on the extent to which Tansley's ecosystem concept was a radical departure from the Clementsian paradigm, with Hagen maintaining that the ecosystem concept has roots in organismal metaphors. Nevertheless, they agree that organismal models of nature have been discredited in ecology.

38. Ibid., 18.

39. Ibid., 32.

40. Worster, *Nature's Economy*, 365.

41. Golley, *A History of the Ecosystem*, 34.

42. Ibid., 201.

43. In this respect, I disagree with Gould (and think he disagrees with himself) when he asserts the moral neutrality of scientific data. What Gould objects to most, and rightly so, is reductionistic and uncritical inferences from scientific information.

44. In *An Entangled Bank*, Hagen contends that Darwin's account contained elements of chaos and lawlessness as well as order and predictability, and as such did not entail a complete break from inherited views.

45. Worster, *Nature's Economy*, 46.

46. This somewhat unfortunate phrase was suggested to Darwin by Herbert Spencer.

Darwin adopted the phrase, even though he would later express surprise that his theory had been popularly (mis)understood to propose that "might makes right."

47. Toulmin, *The Discovery of Time*, 202.

48. Darwin, *Autobiography of Charles Darwin*, 120.

49. It is in this sense that Darwinism is often understood as having broken with teleological explanations of fitness in nature. Gould seems to think it did not break with them sufficiently.

50. The implication of agency in the process of "selecting" is not easy to get around, as Darwin noted: "It is difficult to avoid personifying the word Nature, but I mean by Nature, only the aggregate action and product of many natural laws, and by laws the sequence of events as ascertained by us." Darwin, *The Origin of Species*, 131.

51. Toulmin, *The Discovery of Time*, 227.

52. Darwin, *Autobiography of Charles Darwin*, 90.

53. Ibid., 90.

54. Rachels, *Created from Animals*, 105.

55. Gregersen, "The Cross of Christ in an Evolutionary World, 200."

56. Darwin, *Autobiography of Charles Darwin*, 88.

57. Despite Darwin's skepticism that evolution and Christianity are compatible, some modern Darwinians see the two as not only compatible but even congenial to one another. Philosopher of biology Michael Ruse finds similarities between Darwinism's stress on physical suffering and Christianity's belief that hardship and suffering are an ineradicable part of existence. See Ruse, *Can a Darwinian Be a Christian?*.

58. Worster, *Nature's Economy*, 367.

59. Hagen, *An Entangled Bank*, 129.

60. Worster, *Nature's Economy*, 368.

61. Hagen, *An Entangled Bank*, 148.

62. Golley, *A History of the Ecosystem*, 5–6.

63. Ibid., 5.

64. Hagen, *An Entangled Bank*, 194. Hagen suggests that Daniel Botkin and other nonequilibrium ecologists have exaggerated ecology's earlier commitment to concepts of balance and stability.

65. Worster, *Nature's Economy*, 390.

66. Ibid., 390.

67. Hagen, *An Entangled Bank*, 196.

68. Botkin, *Discordant Harmonies*, 10. It is particularly interesting that ecotheologians associate the idea of nature's balance with an ecological model assumed to have superseded mechanistic models, given that, as Botkin notes, the preoccupation with the idea of balance in early ecology owed much to *machine* models of nature such as those that inspired the Lotka-Volterra equations for predator-prey interaction: "With the machine metaphor for the balance of nature expressed in mathematical models, these became the basis for the [failed] management of fisheries, wildlife, and endangered species. In the social and political movement known as environmentalism, ideas of stability may have been less formal, but the same underlying beliefs of a balance of nature dominated." Ibid., 42. And so they do still among ecotheologians.

69. It is less clear, however, that individual members of a biotic system would experience an outcome that is favorable. While a species as a whole may be "improved" by natural selection, this occurs at the expense of individuals. Improvement itself is also relative—

relative to the specific conditions of the environment at a given point in time. These conditions can change very rapidly and thus improvement has little objective meaning outside of these specific contexts.

70. Golley, *A History of the Ecosystem*, 195.
71. Ibid., 196.
72. Ibid., 196.
73. Worster, *Nature's Economy*, 411.
74. Botkin, *Discordant Harmonies*, 25.
75. Gould, "Darwinism Defined," 70.
76. Gould, *Rocks of Ages*, 6.
77. Rachels, *Created from Animals*, 127.
78. Ibid., 4.
79. Ibid., 126–27.
80. Ibid., 127.
81. Turner, *Reckoning with the Beast*, 61.
82. Ibid., xi.
83. Rachels, *Created from Animals*, 215.
84. Darwin, *The Life and Letters of Charles Darwin*, 3:202–3.
85. Cobbe, *Life of Francis Power Cobbe*, 2:449.
86. Fleming, "Charles Darwin, the Anaesthetic Man," 228.
87. Nash, *The Rights of Nature*, 68.
88. Darwin, *The Descent of Man*, 101.
89. Leopold, *A Sand County Almanac*, 176.
90. Ibid., 203.
91. Callicott, "Can a Theory of Moral Sentiments?"
92. The best-known and most widely cited of these is, again, Lynn White's criticism of anthropocentrism and a linear, progressive notion of time, ideas he attributes largely, but not exclusively, to the "Judeo-Christian" worldview. White, however, also sees the scientific worldview as rooted in this same linear notion of time; as far as natural selection is concerned, his argument is based upon a popular misconception of evolution as inherently progressive.
93. Midgley, "The Paradox of Humanism," 190.
94. Midgley, *Science as Salvation*, 147.
95. Religious myths tend to concentrate more on human sin and salvation, rather than progress and perfection, but humans remain the focus of the narrative nevertheless.
96. Gustafson, *Ethics from a Theocentric Perspective*, 1:88.
97. Ibid., 2:9.
98. Ibid., 2:13.

2. THE BEST OF ALL POSSIBLE WORLDS: ECOFEMINIST VIEWS OF NATURE AND ETHICS

1. Fowler, *The Greening of Protestant Thought*, 153.
2. Ibid.
3. Ruether, *Gaia and God*, 58.
4. Fitzsimmons, "Ecological Confusion Among the Clergy," 2.
5. Ruether, *Gaia and God*, 47.
6. Ruether dispels a survival-of-the-fittest interpretation of evolution. Ibid., 55 ff.

7. Hagen, *An Entangled Bank*, 192.

8. Worster, *Nature's Economy*, 384.

9. Ruether, *Gaia and God*, 4.

10. Lovelock, *The Ages of Gaia*.

11. Lovelock writes, "For me, Gaia is a religious as well as a scientific concept." Ibid., 206.

12. Lovelock and Epton, "In Quest for Gaia," 144.

13. Margulis and Sagan, "God, Gaia, and Biophilia," 351.

14. Ruether, *Gaia and God*, 47.

15. Ibid., 56.

16. Margulis and Sagan, "God, Gaia, and Biophilia," 350.

17. Ibid., 350.

18. Margulis's descriptions of organisms as "murderous" and "rampaging" are far from objective, morally neutral scientific observations. She clearly imposes human ethical categories on nature as much as Ruether does. The primary difference is that Ruether's categories are positive while Margulis's are largely negative.

Ruether is correct to say that organisms competing in nature do not envision the competition as an "enemy" to be annihilated. There is a difference between humans' cultural conception of competition and the competition in nature that results from the fact of resource scarcity. Animals probably do not engage in competition with the *desire* to destroy.

Ruether's point here is well taken. Mary Midgley makes a similar point in criticizing philosophers and scientists who confuse the *fact* of natural competition with a *motive* of competitiveness, thus naturalizing this motive for human societies. Ruether and Midgley share a disdain for reductionist arguments that simply reduce social ethics to "natural" competitive impulses. However, while making a similar argument about competition, they use biological data very differently. Midgley avoids the reductionist claim that human ethics mirrors animals' relations in nature not by denying that competition actually exists in nature but, rather, by denying that the mirroring is absolute and precise. Midgley, in other words, makes Darwinian evolution—in all its competitive and predatory dimensions—central to her account of nature, but she is conservative in drawing ethical norms from this data, arguing that humans inherit evolutionary predispositions and ranges of behavior. Ethics should not consist merely in an inventory of the features of human nature, thereby validating our natural tendencies. Our moral makeup is much more complex than this, Midgley argues.

Ruether takes an opposite approach: Whereas Midgley begins with a more Darwinian view of nature and then draws modest ethical conclusions from it, Ruether deemphasizes Darwinian evolution, focusing on themes of cooperation and mutuality, and then proceeds to derive ethical conclusions liberally from this view. Thus, she argues, we are to consciously emulate nature—a statement Midgley would never make. I will return to Midgley's views in the last two chapters.

19. Margulis and Sagan, "God, Gaia, and Biophilia," 352.

20. Ibid., 352.

21. Ruether, *Gaia and God*, 143.

22. In chapter 6 we will look at two other ways of viewing the role of humans in nature: participation and the idea of management (particularly in terms of wildlife management) that I believe is consistent with it. As I argue there, I do not think these concepts carry the a connotation of "controlling" nature. Guardianship, while not the same as con-

trol, seems in any case to imply something different than the ecological model, as defined by Ruether, suggests.

23. Ruether, *Gaia and God*, 139. Ruether argues here that "although we may evaluate our mortality as tragic, or seek to embrace it as natural, what mortality is not is sin, or the fruit of sin. The (pre-apocalyptic) Hebrew view that mortality is our natural condition, which we share with all other earth beings, and that redemption is the fullness of life within these limits, is a more authentic ethic for ecological living."

24. At one point Ruether acknowledges that the passage from Isaiah regarding peaceful relations among animals "stumbles on the fact that at least a major group of animals also eat other animals." Ibid., 224. This is a point that calls for much greater discussion in her arguments about nature and its moral import.

25. McFague, *Models of God.*

26. Ibid., 93–95.

27. McFague, *Super, Natural Christians*, 67.

28. McFague, "Imaging a Theology of Nature."

29. McFague, *Models of God*, 37.

30. McFague, "Imaging a Theology of Nature."

31. McFague, *Life Abundant*, 32.

32. McFague, *Super, Natural Christians*, 7.

33. Ibid., 2

34. McFague, *Life Abundant*, 33.

35. Ibid., 34.

36. McFague, *Super, Natural Christians*, 1.

37. Ibid., 2.

38. In fact, McFague's argument seems to fail on all three counts, in that the ethic generated is in some sense traditional but not particularly "radical," as we will see. It does not seem to require a significant change in our thoughts or actions toward nonwild nature; when applied to wild nature, on the other hand, it would perhaps be radical, but it would also be misplaced and inappropriate in that context. (I assume her three criteria are intended to be fulfilled simultaneously.)

39. McFague, *Super, Natural Christians*, 7.

40. Ibid., 79. Oddly, McFague attributes the "Enlightenment" separation of nature and humans to Descartes. "Mechanistic" nature is also associated with Newton in her arguments, as with many other ecotheologians. The Newtonian view was "deterministic and atomistic, reductionistic and dualistic, with the machine as the prime analogy." Ibid., 20. In general, the Newtonian and Enlightenment views of nature seem to share a mechanistic, utilitarian perspective, distinct from the views of nature that preceded them, according to McFague's understanding of these movements.

41. Ibid., 52.

42. Ibid., 52.

43. What she really has in mind here is postmodern physics which allegedly undermines "dualistic" distinctions like subject and object.

44. Worster, *Nature's Economy*, 82.

45. McFague, *Super, Natural Christians*, 9.

46. Ibid., 167.

47. Northcott, *The Environment and Christian Ethics*, 160.

48. McFague, *Super, Natural Christians*, 135.

49. Ibid., 9.

50. In fact, it seems that the ecological model has more to do with modern physics than with ecology per se. The idea of radical relationality at first sounds Darwinian, but, as we will see, some ecotheologians appear to derive their "ecological" insights from ideas in field theory, relativity theory, or quantum mechanics. Hence, the strange argument, common in ecotheology, that the "ecological model" is *replacing* the "Newtonian model" (as though biological perspectives have somehow corrected theories in physics).

51. McFague, *Super, Natural Christians*, 21.

52. For example, McFague writes that the "ecological model of self and world is derived from the workings of natural systems: we are interpreting the human self's relation to the world as one instance of ecological interrelationship and interdependence." Ibid., 107.

53. McFague, "Imaging a Theology of Nature."

54. Ibid., 206. McFague uses "metaphors" and "models" somewhat interchangeably. A model, she writes, is a metaphor that has "staying power"—one that "has gained sufficient stability and scope so as to present a pattern for relatively comprehensive and coherent explanation." Ibid., 207.

55. Ibid.

56. Ibid. (my emphasis).

57. McFague, *Models of God*, 36.

58. McFague cannot mean to say that there is *no* difference between the way that metaphors and models work in theology and science (which would indeed be a radical claim). If she did, there would be little reason to pay attention to the "genuine and accurate information" of science that replaces "ignorance and half-truths." McFague, *Super, Natural Christians*, 135. Why should metaphors for God strive to be compatible with scientific knowledge, as she claims, rather than vice versa, if no important differences exist?

In *Metaphorical Theology* McFague writes that "theology is more dependent on its models and fewer opportunities for testing their isomorphisms are available. Certainly in science models can be and are tested by investigating the properties one expects to be present and if the expectations prove to be correct, one concludes that the model *is* a description, albeit imperfect and partial." McFague, *Metaphorical Theology*. A model is different from a theory in science, she argues, in that the former are "possible explanations" while theories are "accepted explanations of phenomena." Ibid., 99. Thus, she seems to believe, science does *aim* at description, while models in theology do not. Moreover, she asserts that "while theories dominate more than models in science, the reverse is the case in theology." Ibid., 103. Whereas science sometimes relies heavily on models and sometimes does not, a theologian "does not have the luxury of deciding between models and no models." In other words, theology must always rely on models: "the question is, which models?" Ibid., 105. Here we return to McFague's criteria for choice of models, one of which, as we have seen, involves compatibility with science. Hence, scientific reality has a different ontological status than the models derived from it, which is why she argues (as she does in *Models of God*) that "we must understand God and the activity of God in the world in a fashion that is not just commensurate with an ecological, evolutionary sensibility but intrinsic to it," rather than suggesting that our understanding of ecology be based upon our understanding of God. McFague, *Metaphorical Theology*, 80. My contention, then, is that she has misconstrued the contemporary scientific picture of reality by neglecting evolution and that her model is problematic.

The question remains, however, whether McFague conceives of the ecological model as a model in *science* or a model in *theology*. It sounds like a model in science, yet she also interprets it as a way of relating to all things, including *God*. McFague claims that models in science deal with *quantitative* dimensions while theological models deal in *qualitative* aspects of life. Ibid., 106. A theological model affects "feelings and actions in the world" and determines "how we conduct ourselves in it." Ibid., 107. Thus the ecological model seems to be a theological, qualitative one; this is reinforced by McFague's interchangeable use of "ecological" and "subjects-subjects" model; see McFague, *Super, Natural, Christians*, 107. If so, it ought still to be compatible with scientific pictures of reality, which, again, I contend that it is not.

Even Ian Barbour, whose work in metaphors and models significantly influenced McFague's theology, does not maintain that models work in precisely the same way in science and religion. Barbour defends a position of "critical realism" that takes models in science and religion "seriously but not literally." Even so, he notes that science accepts *complementary* models for the same entities (e.g., the wave/particle paradox in physics) but *not* the vast multiplicity of models that McFague describes, where mutually exclusive models are permitted in theology because they are not intended to be accurate descriptions. See Barbour, *Religion and Science*, 119. McFague seems to allow a multiplicity of mutually exclusive models as a feature of both science and theology.

Without getting too deeply involved in this complicated issue, I would argue that metaphor works differently in theology, which makes claims beyond those involved in ordinary human ways of knowing. I agree with Barbour's critical realism in science but think that less realism (or perhaps a *more* critical version of it) is warranted in theology. Models in theology, such as models of God, persist in a different way from models in science, which typically are *either* superseded *or* continue to hold sway (this, as I understand it, is part of Barbour's point about complementarity and multiplicity of models). Regarding complementarity in science, "the use of one model limits the use of other models," and the acceptance of complementarity "cannot be used to avoid dealing with inconsistencies." Barbour, *Religion and Science*, 169–70. It is also likely that models play a different role in different kinds of sciences, and it is not clear to me that biology depends upon models in the same way that, say, physics does, since the latter deals more in unobservables. Thus talk about black holes and talk about God *may* involve similar reliance on models.

59. McFague, *Super, Natural Christians*, 152.

60. McFague, *Models of God*, 11.

61. McFague, *Super, Natural Christians*, 155.

62. Ibid., 105.

63. Ibid., 158 (my emphasis).

64. Ibid., 162.

65. This notion of the ecological self is a theme of *Life Abundant* as well. See McFague, *Life Abundant*, pp. 31–33.

66. McFague, *Super, Natural Christians*, 154.

67. McFague seems to give some preference to places where humans and nature intersect and are linked in their oppression: "Christians' special focus might be on the shared earth of our cities, where people, often poor ones, and nature, often deteriorating land, meet." McFague, *Super, Natural Christians*, 4.

68. This is why "survival of the fittest" is a misleading phrase. Reproduction, rather than survival, is what is significant for natural selection. Organisms need not survive long after reproducing in order to meet the criterion of Darwinian "fitness."

69. McFague, *Life Abundant*, 36.

70. Ibid., 142.

71. Ibid., 36.

72. McFague, *Super, Natural Christians,* 15.

73. Ibid.

74. McFague acknowledges at one point that for "some to be in 'perfect health,' given our limited resources, means for others to die." McFague, *Models of God,* 148. This is an important point that demands greater emphasis than it receives in her work. Furthermore, this recognition does not deter McFague from her basic commitment to attending to the bodily needs of all individual beings in the world—an objective that runs counter to evolutionary processes. McFague, in other words, briefly acknowledges the inevitability of suffering and death but merely interprets this natural fact as a confirmation of difficulty of Christianity's struggle against "evil" and "sin" in the world. Ibid. She does not realize that it is the "evolutionary, ecological" process *itself* against which she is struggling.

75. Rolston, *Conserving Natural Value*, 67.

76. Northcott, *The Environment and Christian Ethics*, 161. Whereas Northcott sees McFague, and many ecotheologians, as banning all hierarchies in favor of a "homogenizing" ethic that falsely elides differences, I would also point out that the homogenizing move unwittingly retains hierarchical distinctions: animals are like us and those most like us are worth more. The move is at once homogenizing and (inappropriately) hierarchical.

77. McFague, *Super, Natural Christians*, 158.

78. Ibid., 166.

79. Thus it appears once again that the ecological model is a *theological* model, not a scientific one. Yet even as a theological model, it is not a particularly appropriate one.

80. Or, again, if it is applied to nonwild nature, it would not generate behavior that could be considered a radical departure from our current approach.

81. Northcott, *The Environment and Christian Ethics*, 117.

3. THE ECOLOGICAL MODEL AND THE REANIMATION OF NATURE

1. Moltmann, *God in Creation*, 2.

2. Ibid., 11.

3. Ibid., 17.

4. Ibid., 12.

5. Fowler, *The Greening of Protestant Thought*, 178.

6. Moltmann, *God in Creation*, 2.

7. Ibid., 3.

8. Ibid., 55.

9. Ibid., 66.

10. Moltmann's emphasis on biblical cycles of rest resembles Ruether's account of the covenant metaphor as implying guidelines for our treatment of nature.

11. Moltmann, *God in Creation*, 287.

12. Worster, *Nature's Economy*, 35.

13. Ibid., 35.

14. Moltmann, *God in Creation*, 148.

15. Descartes, *Discourse on Method*, 1:115–18.

16. Bowler, *Evolution*, 253.

17. Ibid., 211.

18. Ibid.

19. This is not to say that there are no other alternatives for reconciling evolution and belief in God, but that those alternatives are often less attractive. Something must be given up: either the traditional understanding of God must be altered or the processes of evolution must be reinterpreted along less Darwinian lines. Process theology, for instance, manages a kind of reconciliation by denying some of the traditional attributes of God, specifically, the idea that God is all-powerful or that he compels nature in various directions. A common corollary is that God suffers along with suffering creation (an idea found in McFague's theology as well as some process theology).

20. Moltmann, *God in Creation*, 100 (my emphasis).

21. "Creation" is one way of interpreting nature. Process thinkers refer to creation but also attempt to understand the natural world in scientifically accurate ways.

22. Birch and Cobb, *The Liberation of Life*.

23. Goodwin, *How the Leopard Changed Its Spots*, 191.

24. There are still a number of difficulties with Goodwin's argument and his discussion shifts back and forth between nature as a whole as subject and the value of individual "subjects" with what he calls "first person, subjective" consciousness. But on the whole, he does not focus solely on these individual subjects; moreover, Goodwin tries to integrate the mechanical and organismic models of life rather than simply rejecting the former.

25. Goodwin, *How the Leopard Changed Its Spots*, 215.

26. Cobb and Griffin, *Process Theology*, 29.

27. Ibid., 73–74.

28. Birch is trained as a biologist.

29. Cobb and Griffin, *Process Theology*, 21.

30. Ibid., 154.

31. Ibid., 24.

32. Ibid., 67.

33. Birch and Cobb, *The Liberation of Life*, 89.

34. Ibid., 135.

35. Ibid., 135.

36. Note the similarity here with Augustine's account of animal and human life, in keeping with his basic distinction between earthly peace and heavenly peace, or peace of the body and peace of the soul. "If we were irrational animals, our only aim would be the adjustments of the parts of the body in due proportion, and the quieting of the appetites—only, that is, the repose of the flesh, and an adequate supply of pleasures, so that bodily peace might promote the peace of the soul. . . . But because there is in man a rational soul, he subordinates to the peace of the rational soul all that part of his nature which he shares with the beasts, so that he may engage in deliberate thought and act in accordance with his thought, so that he may thus exhibit that ordered agreement of cognition and action which we called the peace of the rational soul." Birch and Cobb describe a similar quality in humans in our "disciplining of our bodies to put up with some weariness and some discomfort because of concerns that belong to our human experience. . . . As long as the animal experience functioned in accordance with the needs of the body, there was unbroken harmony, they argue, "but now we are subject to a force that restricts us from within." Birch and Cobb, *The Liberation of Life*, 135–36. In Augustine's account, too, the human subdues the body by means of an inner force—the will—that helps order our lives. Human nature, in consisting of more than a peace or "harmony" of the body, involves greater possibilities for discord, for internal struggle. We are subjected, Augustine argues, to the "distress of

pain and grief, the disturbance of desire, the dissolution of death" and seek a different sort of harmony than animals. Augustine, *City of God*, 872–78.

37. Here Birch and Cobb begin to sound like Moltmann in his account of animal environments and human environments.

38. Cobb and Griffin, *Process Theology*, 69.

39. Birch and Cobb, *The Liberation of Life*, 109. It is not always clear whether Birch and Cobb are proposing a qualitative difference between human and animal experience or a quantitative one: for instance, humans are described as having "more" richness of experience than animals, but they also seem to suggest that our experience is fundamentally different.

40. Ibid., 109.

41. Ibid., 131.

42. Ibid., 107.

43. Cobb and Griffin, *Process Theology*, 119.

44. Ibid., 119.

45. They maintain that the nonanthropocentric nature of their argument lies in its treatment of all beings as (to some degree) both ends and means. In other words, because they do *not* treat humans as absolute ends in themselves, never to be regarded as means also, they believe that their approach avoids anthropocentrism. Still, humans are "primarily ends" rather than means, Birch and Cobb argue. Birch and Cobb, *The Liberation of Life*, 162. They do not draw absolute distinctions between humans and other animals; nevertheless, they evaluate all other animals in light of human traits and experiences. I will return to a more detailed discussion of the anthropocentrism of Birch and Cobb's approach in chapter 5.

46. Ibid., 158.

47. Their use of rights terminology is not well defined. Rights correspond to richness of experience and thus we can assume that some animals have a (negative) "right" not to be made to suffer. Other than this, it is not clear what sorts of things animals have a *right to* (they deny that animals have absolute rights to life). Some animals, such as chickens, who do not have a well-developed social life or an ability to anticipate death or grieve the death of others, do not possess any robust kinds of rights. Birch and Cobb do not offer a coherent account of rights nor do they consistently use this language.

48. Birch and Cobb, *The Liberation of Life*, 155.

49. Rolston, *Environmental Ethics*, 54.

50. Birch, "Christian Obligation," 58.

51. Birch understands the fact of physiological continuity between organisms (for instance, the evolution of a central nervous system in different animals) to be an evolutionary basis for his ethic of reducing suffering among certain species.

52. Birch "Christian Obligation," 68.

53. Ibid., 67.

54. "Plea to Cherish Mini-Beasts," *BBC News*, online, June 5, 2000, available at http://news.bbc.co.uk/hi/english/sci/tech/newsid_778000/778365.stm.

55. Ibid.

56. Birch and Cobb's claim that animals without richness of experience still have instrumental value begs for elaboration. They suggest that such organisms (invertebrates and plants) have great instrumental value because of the other life-forms that depend upon them. This is closer to an ecological analysis than many of their other arguments about

value. Yet they never articulate what, if any, obligations we have toward these lesser life-forms. The argument that we have obligations primarily to those beings that can suffer discounts numerous life-forms whose suffering is minimal (because of their lack of a central nervous system) or nonexistent (such as plants). Birch clearly associates intensity of feeling with richness of experience and, in turn, assigns greater moral obligation to those organisms with richness: "Intrinsic value in creatures like ourselves makes an ethical claim upon us to recognize our obligation toward them." Birch, "Christian Obligation," 60.

57. Nixon, "The Species Only a Mother Could Love."

58. Ibid., 28–29.

59. Cobb and Griffin, *Process Theology*, 76–77. Process thinkers might respond that we have an obligation to preserve the mussels for the sake of those higher beings that do have experience and can enjoy the mussels (as food, for instance.) Such a claim would be less anthropocentric. But the basic emphasis upon experience still gives no consideration to other important ecological issues such as whether or not the species in question are native or introduced. What happens in cases of direct conflict, when environmentalists must choose between preserving a native, nonsentient, nonexperiencing animal and an experiencing, introduced one? (We will look at such a case in chapter 5.) Or how are decisions made when humans and animals compete for use of a species that has instrumental value to both (as in the case of elephants and humans below)? We might try to strike some compromise such that all experiencing species can "use" this nonexperiencing species but, in the end, humans are given ultimate access to resources due to their unique value as primarily ends in themselves.

60. Rolston, "Environmental Ethics in Antarctica." Rolston points out that even land ethics ceases to be very useful in such a situation, since in Antarctica there is little in the way of an ecosystem in the sense that scientists usually understand it.

61. Ibid., 132. For practical reasons, too, duties are intensified because nature is even more fragile and vulnerable in a place such as Antarctica. Humans only visit those few places where there are plants or animals to observe and whatever waste they leave behind takes far longer to decompose in the frigid climate than in other places.

62. Birch and Cobb, *The Liberation of Life*, 161.

63. Ibid.

64. Ibid., 173.

65. Ibid., 162.

66. Cobb and Griffin, *Process Theology*, 73.

67. Birch and Cobb, *The Liberation of Life*, 193.

68. Rasmussen, *Earth Community*, 155.

69. This is not to say that animals do not ever respond intelligently or creatively to their environments or that they cannot learn by experience. Genetic adaptations are sometimes preceded by nongenetic shifts in behavior. Certain genetic and morphological changes may subsequently be favored by natural selection because of new habits and behaviors. Animals' ability to learn can have an effect on changes in selection pressures. Behavior may set up new selective pressures that would preserve a random variation that codes for certain features. But evolution (genetic variations and morphological changes) is not provoked directly by the new environment. Richard Lewontin's "dialectical" view of organisms, discussed below, grasps this more complex account. Birch and Cobb mistakenly assign a central role to the creative, intelligent responses of organisms to their environments and interpret this as the main "force" in evolution. Moreover, if evolution occurred in this fashion, provoking adaptive responses from organisms, there would be far less

struggle and competition in nature than Darwinians believe. In assuming that it is the organisms' *response* that introduces novelty, they discount the novelty that is contributed through random variation or, perhaps, they believe that variation and response are one and the same—i.e., variation directed toward adaptation, as Lamarckian theories hold.

70. Bowler, *Evolution*, 248.

71. Ibid., 248.

72. Numbers, *Darwinism Comes to America*, 35.

73. Bowler, *The Non-Darwinian Revolution*, 100.

74. Birch and Cobb, *The Liberation of Life*, 197.

75. Ibid., 195.

76. Stephen Jay Gould, "Shades of Lamarck," in Gould, *The Panda's Thumb*, 77.

77. Ibid., 81.

78. Here we can see why Birch and Cobb believe that their ecological model is best exemplified by humans, who face new challenges and adversity, and in responding to these, continually propel evolution (at least as cultural evolution) forward.

79. Birch and Cobb, *The Liberation of Life*, 196.

80. Richard Levins and Richard Lewontin, "Organism as Subject and Object," in Levins and Lewontin, *The Dialectical Biologist*, 86.

81. Ibid., 86–88.

82. Ibid., 89, 106.

83. Just as ecotheology's vindication of the oppressed in nature revolts against a Social Darwinist view that might makes right, its vindication of organisms as subjects may also be a reaction against a particular ("objectifying") reading of Darwinism that is not necessarily accurate. The upshot is that a more informed understanding of Darwinism might also make the theory somewhat more palatable to ecotheologians.

4. DARWINIAN EQUALITY FOR ALL: SECULAR VIEWS OF ANIMAL RIGHTS AND LIBERATION

1. In chapter 6, I explore this distinction between different kinds of interrelationship and interdependence in greater detail.

2. Singer, *Practical Ethics*, 85.

3. Whereas secular accounts of animal rights and liberation appear to have emerged in part from civil rights movements (hence, as we will see, the alleged parallels between speciesism and racism), these arguments in religious environmental ethics have as much to do with biblical themes of liberation (found in Exodus and in the teachings of Jesus) as they do with civil rights. I do not know whether any attempts have been made to trace the origins—separate or shared—of secular and religious arguments for animal liberation. In any case, it seems that the two have converged on the argument that similar forms of oppression exist in humans and animals (and liberation theology has both biblical and political/social origins, just as the civil rights movement invokes biblical symbols and stories in support of human rights). The gradual inclusion of *animal* liberation in both kinds of arguments is interesting.

4. Indeed, Christian environmentalists ought perhaps to be even more cautious about issuing claims regarding animal rights, liberation, and suffering, in that they hope to extend their views to "nature" and "creation" at large, or "life" in general, whereas Singer and Regan are generally more modest about the classes of organisms to which their arguments are meant to apply. In general they are more aware of the difficulties of extending certain

claims to the entire spectrum of life and focus much, though not all, of their attention on nonwild animals. The broader focus of ecotheologians on all living things makes their claims even more implausible.

5. Singer, *Animal Liberation*, 8. Occasionally, Singer does use the term *rights* (more so in his *Practical Ethics* than in *Animal Liberation*), but the concept of rights is not central to his argument.

6. See Ryder, *Victims of Science*.

7. A version of this argument is articulated by Moltmann, as we have seen, in his view of humans as God's proxy on earth.

8. Ryder, *Victims of Science*, 5.

9. Singer has been called "a public advocate of genocide and the most dangerous man on earth" by one such critic, Diane Coleman, whose group, Not Dead Yet, has protested his Princeton appointment. Quoted in Specter, "The Dangerous Philosopher," *New Yorker*, Sept. 6, 1999, 52.

10. Singer, *Animal Liberation*, iv.

11. Birch and Cobb, *The Liberation of Life*, 2.

12. McFague, *Super, Natural Christians*, 170.

13. Singer, *Animal Liberation*, 20–21.

14. Ibid., 20.

15. Ibid.

16. Ibid., 7.

17. Bentham, *An Introduction to the Principles*.

18. Singer, *Animal Liberation*, 17.

19. Singer, *Practical Ethics*, 67.

20. Ibid., 67.

21. Ibid. Inflicting pain and killing are obviously connected here. Considerations of mental level and other capacities become important when the issue is taking life.

22. Singer, *Animal Liberation*, 21.

23. Ibid., 15.

24. Ibid., 17.

25. Singer, *Practical Ethics*, 57.

26. Disabled persons and advocates of their rights take particular offense at Singer's argument that some humans do not qualify as persons and have been vocal participants in protests of his Princeton appointment. Singer would surely point out that anyone who is capable of the act of protesting, and of defending his or her rights—verbally or otherwise—already qualifies as a person.

27. Singer briefly discusses Joseph Fletcher's account of being "human" in *Practical Ethics*, 86–87.

28. Ibid., 119.

29. Ibid., 118.

30. Ibid., 169. However, an infant's life may be of great value to others, particularly parents, thus Singer often refers to *orphans* in drawing comparisons between the wrongness of killing different kinds of beings.

31. Singer, *Animal Liberation*, 17.

32. Ibid., 225.

33. Despite the fact that Singer here takes a hands-off approach to animals in the wild, however, his discussion of person and nonperson animals indicates that these categories

cross the domestic/wild boundary. If, as Singer suggests, we have greater responsibilities toward the lives of persons, such obligation presumably holds regardless of the environment in which animals exist.

34. Ibid., 226.

35. Ibid.

36. Ibid., 227. This argument is similar to Birch and Cobb's claim that avoidance of suffering and discord may result in a life of triviality, devoid of intense experience. Singer's argument here also seems to suggest that pain is not the opposite of pleasure—a life with anxiety, distress, and the risk of bodily harm is also pleasant. It would be difficult to judge whether, on the balance, such an organism had a life that was more "pleasant" than "painful."

In any case, Stephen Budiansky has responded to Singer's claim here about a life of boredom versus freedom. He argues that animals have been known to "choose" confinement, and boredom, so long as they get a steady supply of food. If they always preferred freedom, domestication would not have been possible. See Budiansky, *The Covenant of the Wild*, 148.

37. Singer, *Animal Liberation*, 227.

38. Ibid., 13.

39. Ibid., 11.

40. Ibid.

41. Callicott, "Animal Liberation," 54.

42. Ibid., 56.

43. Callicott, "Animal Liberation and Environmental Ethics," 258. In fairness to Singer, I would say that it is not clear that he formally endorses any such predator-extermination policy. Callicott observes in a footnote that Singer "toys" with the idea; this is accurate, I think, but Singer appears ultimately to reject it, stating that we should not interfere in these natural encounters with predators and that reduction of the suffering of lab and farm animals is sufficient. Callicott's general point, however, is well taken: an ethic of reduced suffering is inapplicable to wild animals and would seem to imply this sort of predator control *if* widely implemented (whether or not Singer explicitly advocates such a policy).

44. Singer, *Practical Ethics*, 271.

45. Ibid., 273.

46. He cites land ethics and other "holistic" approaches, dubbing all of these "deep" ecology approaches, despite the important philosophical differences between land ethics and the movement properly known as deep ecology. All such approaches, he argues, attempt to explain the interests of entities that can, in fact, have no interests.

47. Singer, "Not for Humans Only," 203.

48. Singer, *Practical Ethics*, 284.

49. Admittedly, Singer might counter (as he suggests in *Animal Liberation*) that a wild animal whose physical needs were thus met would suffer a life of such extreme boredom that it would be worse than living in the wild. But, theoretically, even animals' need for a more exciting life could be met without being in the wild.

50. Rolston, "Duties to Endangered Species," 68.

51. Ibid., 66.

52. Ibid., 70.

53. Ibid., 63.

54. Singer, *Practical Ethics*, 95.

55. Ibid., 61.

56. Singer does not use the term *semi-persons*, but this is what I take him to be saying.

57. Ibid., 117.

58. Interestingly, Birch and Cobb also confer great value on dolphins as having an unusually rich subjective life, whereas Singer has some doubt about their degree of intelligence and self-awareness and places them somewhat below apes. These disagreements over precisely how "rich" a mental life—and hence how much value—certain organisms have illustrate why valuing animals in light of such criteria is problematic. What constitutes the most *definitive* signs of intelligence seems a rather subjective matter, particularly when comparing two or more organisms who share the same physiological markers of an advanced mental life.

59. In chapter 5 we will look at several such cases.

60. National Research Council, *Science and the Endangered Species Act*, 9.

61. Air and water are also crucial for our survival, yet there are very few places left in the world where air and water pollution are unknown, and in many places their contamination reaches quite dangerous levels. Environmental arguments based on self-interest do not seem to work as they should.

62. Alberta Agriculture, Food and Rural Development, "Ginseng Industry," http://www.agric.gov.ab.ca/agdex/1000/8883002.html.

63. U.S. Fish and Wildlife Service, "Santa Barbara Island Liveforever (*Dudleya traskiae*) Recovery Plan" (Portland: U.S. Fish and Wildlife Service, 1985).

64. Birch and Cobb, *The Liberation of Life*, 154.

65. Regan, "Animal Rights, Human Wrongs," 41.

66. Regan spells out cases in which individual rights can be overridden in chapter 8 of *The Case for Animal Rights*. I will not pursue these exceptional cases here.

67. Regan's argument here is similar to Bryan Norton's objection that habitat could be destroyed on the grounds that less suffering is entailed overall.

68. Regan, "Animal Rights, Human Wrongs," 39.

69. Regan, *The Case for Animal Rights*, 32.

70. Ibid., 81.

71. Regan, "The Radical Egalitarian Case," 41.

72. Regan, *The Case for Animal Rights*, 262.

73. Ibid., 262.

74. Ibid., 100.

75. Ibid., 262.

76. Regan, "Christianity and Animal Rights," 74.

Regan's argument here is questionable. Likening his position on animal rights to cases of murder, rape, and discrimination confuses the issue somewhat, in that a category such as "murder" is wrong by definition. Murder is a species of the larger category of killing. To say that killing is always and everywhere wrong might be an extreme position, but it hardly seems extreme to take a consistent ethical stand against murder.

77. Regan, "The Radical Egalitarian Case," 41.

78. Regan, *The Case for Animal Rights*, 366.

79. Ibid.

80. Ibid.

81. Regan, "The Radical Egalitarian Case," 45.

82. Ibid., 45

83. Regan, "Christianity and Animal Rights," 79.

84. Singer, *Animal Liberation*, 85.

85. Regan, "Christianity and Animal Rights," 80.

86. Singer acknowledges that the Bible contains "scattered passages in the Old Testament" encouraging something along the lines of kindness toward animals and possibly promoting stewardship rather than outright tyranny and cruelty. However, even these present "no serious challenge to the overall view, laid down in Genesis, that the human species is the pinnacle of creation and has God's permission to kill and eat other animals." Singer, *Animal Liberation*, 188.

87. Regan, "Christianity and Animal Rights," 77.

88. Regan, *The Case for Animal Rights*, 361.

89. Ibid., 361–62.

90. Ibid. Regan interprets land ethics in several ways that are mistaken. He apparently understands it as a crude utilitarian measurement of what "maximizes" the "beauty, stability, and integrity" of a biotic community. Second, he assumes that humans are just another part of this calculus—plentiful individuals who do not contribute to the biotic community, whose numbers cause more harm than good, and can thus be sacrificed. (This, of course, would be an odd kind of utilitarianism that would sacrifice the many to preserve the few, but Regan seems to believe that land ethics, like utilitarianism, counts the greater good of the biotic community over the rights of individuals.) However, Regan does not understand that it is not *sheer numbers* that determine how decisions are made in land ethics (more on this later). Furthermore, neither Callicott nor Rolston treat humans as just another over-represented species that threatens ecosystems.

91. Ibid., 363.

92. Someone has developed an argument along these lines, at least from a legal standpoint. See Christopher Stone, *Should Trees Have Standing? Toward Legal Rights for Natural Objects* (Los Altos: William Kaufman), 1974.

93. Regan, *The Case for Animal Rights*, 363.

94. Ibid., 246.

95. Ibid., 363.

96. Under the Endangered Species Act, "subspecies" and "distinct population segments" also sometimes qualify for protection.

97. Rolston, "Duties to Endangered Species," 70.

98. Regan, *The Case for Animal Rights*, 359.

99. Ibid., 360.

100. One philosopher who deals more adequately with the issue of harm and respect for organisms is Paul Taylor in *Respect for Nature*. Alongside principles such as nonmaleficence and noninterference, Taylor's spells out a concept of "restitutive justice" for animals who have already suffered harms. This principle ensures that a "balance of justice between a moral agent and a moral subject when the subject has been wronged by the agent." Restitutive justice "requires that the agent make reparation by returning those organisms to a condition in which they can pursue their good as well as they did before the injustice was done." Taylor, *Respect for Nature*, 187–88.

101. Regan, *The Case for Animal Rights*, 94.

102. Even if Regan would continue to object to the idea of endangered *species* as worthy of protection (as opposed to individuals), their path to endangerment involved harm to individual animals. Thus it is hard to understand how Regan could not care about the

process of endangerment as a whole, regardless of whether the focus is on the individual or the species as a unit. Rolston, for instance, would not deny that extinction involves harm to individuals but only that the continuation of the species line is equivalent to the continuation of each individual. In other words, he would concede that speciation is not *necessarily* good for both the species and each individual (e.g., as in the case of genetic load), but actions that lead to the extinction of a species (poaching, habitat destruction, etc.) can certainly be bad for individuals as well as the species as a whole.

103. Rolston, "Duties to Endangered Species," 72–73 (my emphasis).

104. Ibid., 72.

105. Sagoff, "Animal Liberation and Environmental Ethics," 87.

106. For instance, one of the chief virtues of the human genome project, researchers often argue, is its potential to undermine race-based discrimination by showing that it has no genetic foundation.

107. Midgley, "The Significance of Species," 124.

108. Rolston, "Duties to Endangered Species," 65.

109. Midgley acknowledges that the issue is more complex than this as far as human "racial" ethics are concerned; diversity itself is a good to be protected and encouraged and members of racial minorities do not necessarily want to be treated as though their race is completely irrelevant. But deciding that we have no moral obligations toward certain people because of their race is obviously unethical.

110. Midgley, "The Significance of Species," 122.

111. Ibid., 124.

112. Ibid., 130.

113. Stephen Budiansky has argued persuasively that domestication of animals is not exploitation but rather a mutually beneficial evolutionary strategy. He argues that we need to stop looking at domestication as only a human phenomenon and consider "what's in it for [animals]." Budiansky, *The Covenant of the Wild*, 41.

114. We may legitimately give preference to certain nonhuman animals over others, not because of (or in spite of) their species membership but because of our particular bonds with them: Midgley argues, for example, that we have greater obligations to our own pets than we do toward other animals. Thus, a serious problem with Singer's argument, in her view, is that he "allows a principle of selection based on the varying nervous capacities of different animals, but refuses to supplement it by a further principle based on nearness or kinship." Midgley, *Animals and Why They Matter*, 96. Some animals make *social* claims upon us, while others, such as wild species, make claims that are primarily *ecological* in nature.

5. PHILOSOPHICAL AND THEOLOGICAL CRITIQUES OF ECOLOGICAL THEOLOGY: BROADENING ENVIRONMENTAL ETHICS FROM ECOCENTRIC AND THEOCENTRIC PERSPECTIVES

1. Leopold, *A Sand County Almanac*, vii.

2. One of Leopold's biographers, Marybeth Lorbiecki, writes: "Putting his head on a clump of grass, he folded his arms over his chest. . . . Leopold died of a heart attack before anyone knew he was in trouble. The fire passed lightly over him." Lorbiecki, *A Fierce Green Fire*, 179.

3. Leopold's writing constantly blends organismic and mechanical metaphors.

4. Leopold, *A Sand County Almanac*, 170.

5. Ibid., 132. As Daniel Botkin observes, Leopold's argument that elimination of predators resulted in an unnatural, sharp rise in deer herds relied on an overly simplistic assumption of a "balanced" relationship between predator and prey. Botkin argues that the role of predators is in fact less clear; deer population levels "might be the result of a reduction in competition rather than a decrease in predation." Botkin, *Discordant Harmonies*, 78. However, as Botkin concludes (and as I wish to argue), environmentalists need not abandon these ideas altogether nor dispense with Leopold's land ethic. Rather, we need to reject a naive interpretation of ecological stability and harmony, embracing a "redefined" account of nature as discordant harmony—a view, as Botkin asserts, that is "also consistent with the land ethic of Aldo Leopold." Ibid., 191.

6. Leopold, *A Sand County Almanac*, 130.

7. Robert Finch, "Foreword," ibid., xvii.

8. Leopold, *A Sand County Almanac*, 225.

9. Ibid., 204.

10. Callicott, "The Conceptual Foundations," 230.

11. Leopold, *A Sand County Almanac*,, 109.

12. Ibid., 110.

13. See, for example, ibid., 211, where Leopold argues that birds have a "biotic right."

14. Leopold, "Some Fundamentals of Conservation in the Southwest," 140.

15. Leopold, *A Sand County Almanac*, 214.

16. Ibid., 215.

17. At least one ecotheologian has defended the view that predator-prey interaction—eating and being eaten—reflects a divine ordering and ought to be received with a sense of gratitude. Matthew Fox argues that "one of the laws of the universe is that we all eat and get eaten. In fact, I call this the Eucharistic law of the Universe." Fox, "Green Spirituality."

18. Leopold, *A Sand County Almanac*, 214–15.

19. Rasmussen, *Earth Community*, 347.

20. Callicott, "The Conceptual Foundations," 234. Note that this statement illustrates Regan's mistaken impression of land ethics as treating all forms of life, human and nonhuman alike, as subject to the same ethic. Callicott clearly recognizes that there are different moral "codes" in human society and in wild nature.

21. Ibid., 231.

22. Rolston, *Environmental Ethics*, 161.

23. Leopold, *A Sand County Almanac*, 215.

24. Ibid., 7.

25. Rolston, *Environmental Ethics*, 175.

26. Ibid., 175.

27. Ibid., 223.

28. Callicott in particular has devoted much of his career to defending and fine-tuning the land ethic. See, for example, his *In Defense of the Land Ethic*.

29. On the issue of wilderness, see William Cronon's important essay problematizing the idea of wilderness, "The Trouble With Wilderness." Cronon criticizes the "habits of thinking" that emerge from the "complex cultural construction" that environmentalists refer to as wilderness. Cronon is correct in his critique of the romanticizing tendency of much of environmentalism. I agree, too, with Cronon's assertion that wilderness is a place that forces us to acknowledge the existence of forms of life "not of our making," the interests of which are not necessarily identical to our own. The value of nature is intimately

linked to this acknowledgment, in my view—and in the view of theocentric ethics. At the same time, I would argue that in staking out his own middle-ground position, Cronon hyperextends the difference between two poles in environmental thinking, claiming that environmentalists draw a sharp line between cultivated, human-corrupted nature, on the one hand, and wilderness as pure, pristine, and devoid of traces of human existence, on the other. As examples of those who value only the unfallen, edenic, remote parts of nature Cronon cites somewhat "extreme" environmental movements such as deep ecology and Earthfirst! In doing so, he overlooks an already existing middle ground between these two poles, a middle ground that many environmentalists in fact occupy.

30. Callicott's position on whether or not pain in nature is a "good" or "bad" thing seems to have changed somewhat over time. His view has been that the goodness or badness of pain is relative to the functioning of an ecosystem—that it is instrumentally good or bad. See Callicott, "Animal Liberation." However, he seems to have modified his views on this issue (and others related to it, such as vegetarianism) in some of his subsequent work, e.g., "Animal Liberation and Environmental Ethics."

31. See Callicott, "Rolston on Intrinsic Value."

32. For more detailed discussion of Rolston's account of systemic, nonanthropogenic disvalues, and the theological implications thereof, see Attfield, "Evolution, Theodicy, and Value."

33. Rolston, "Disvalues in Nature."

34. Gregersen, "The Cross of Christ in an Evolutionary World," 201.

35. What I see as a central virtue in Rolston's environmental ethic—the ethical primacy of species over individuals in nature—others see as a weakness in his account of "evil." Jill Le Blanc argues that Rolston answers (or sidesteps) questions of theodicy by simply shifting attention to abstract, systemic ultimate goods, while forgetting the "suffering of a single individual animal starving." Le Blanc, "A Mystical Response to Disvalue in Nature," 259.

36. Attfield, "Evolution, Theodicy, and Value," 294.

37. Callicott, "The Wilderness Idea Revisited," 348.

38. Rolston, "The Wilderness Idea Reaffirmed."

In citing "cultural" values, Rolston means that Callicott prizes land too often for its value for humans. I tend to side with Rolston in this debate about nature and culture.

39. Ibid., 368.

40. Ibid.

41. Rolston, "Challenges in Environmental Ethics," 138.

42. Rolston, *Conserving Natural Value*, 112

43. Ibid., 113.

44. Ibid.

45. Rolston, "Biology and Philosophy in Yellowstone," 245.

46. In *Playing God in Yellowstone*, Alston Chase argues that wildlife managers have elevated natural management guidelines to the status of philosophical and religious dogma, thereby "destroying" the park.

47. It is worth noting that Rolston does not condemn the protests or actions of the Fund for Animals in this case, even though he applauds the park's decision (which was later upheld by the courts). He writes that "happily," some of the goats were removed rather than shot but "unhappily," it is not possible to rescue and move all of them. The death of the goats is tragic, even though the action is justifiable. Rolston, "Duties Toward Endangered Species," 67.

48. For this reason, I am not certain that a theory of moral sentiments, in and of itself, could sufficiently guide environmental ethics.

49. Rolston's works, including *Environmental Ethics* and *Conserving Natural Value*, discuss a far wider range of values than those I will attempt to examine here.

50. Rolston, *Conserving Natural Value*, 42.

51. Rolston's account of attraction to charismatic animals illustrates that his ethic is not exclusively deontological, though other ethicists (such as Northcott for example) categorize his ethics as such.

52. National Research Council, *Science and the Endangered Species Act*, 112.

53. Birch and Cobb, *The Liberation of Life*, 155.

54. Rolston, *Conserving Natural Value*, 104.

55. Rolston, "Challenges in Environmental Ethics," 136.

56. Rolston, *Conserving Natural Value*, 59.

57. Rolston, *Environmental Ethics*.

58. Rolston, *Conserving Natural Value*, 61.

59. Rolston, *Environmental Ethics*, 188.

60. Rolston, *Conserving Natural Value*, 61.

61. Rolston, *Environmental Ethics*, 190.

62. Rolston, *Conserving Natural Value*, 60.

63. Ibid.

64. Birch and Cobb, *The Liberation of Life*, 88.

65. Wynn, "Natural Theology in an Ecological Mode," 39.

66. See Rolston's discussion of duties to "sentient life" in chapter 2 of *Environmental Ethics*.

67. Ibid., 191.

68. This is the argument presented very persuasively by Bill McKibben in *The End of Nature*. McKibben however would put the significant turning point of our alteration of the environment in the second half of the twentieth century, with the advent of global warming and ozone depletion, which are a dramatic departure from all previous forms of anthropogenic pollution.

69. McFague, *Super, Natural Christians*, 158.

70. Birch, "Christian Obligation," 68.

71. Ibid. (my emphasis).

72. Moltmann, *God in Creation*, 56.

73. Ibid., 12.

74. Rolston, "Wildlife and Wildlands," 135.

75. Rolston, *Environmental Ethics*, 86.

76. Rolston, "Wildlife and Wildlands," 135.

77. Rolston, "Challenges in Environmental Ethics," 140.

78. Rolston, "Wildlife and Wildlands," 135.

79. Callicott, "Animal Liberation," 32.

80. Ibid., 34.

81. Ibid., 37.

82. Ibid., 54.

83. Rolston, *Conserving Natural Value*, 69.

84. Ibid., 70.

85. Ibid.

86. Leopold, *For the Health of the Land*, 22.

87. Aldo Leopold, "The Land Health Concept," 221.

88. Leopold, "The Land Health Concept," 219. Leopold considered pest species of animals to be similar to weeds—that is, species that, because of their introduction to an area or because of the loss of their predators, come to overrun the land. Attributes such as "pest" or "good" and "bad," he wrote, "are attributes of numbers, not of species." Leopold, "What Is a Weed?" 212.

89. Callicott, "Do Deconstructive Ecology and Sociobiology?" 369–72.

90. Ibid., 369.

91. Leopold, *A Sand County Almanac*, 132.

92. This essentially is Joel Hagen's conclusion in *An Entangled Bank*.

93. Callicott, "Do Deconstructive Ecology and Sociobiology?" 371.

94. Ibid.

95. Ibid.

96. Leopold too wrote that *wildlands* designates a place retaining "representative samples" of "original or wilderness condition, to serve science as a sample of normality." Leopold, "Planning for Wildlife," 179.

97. Rolston, "Does Nature Need to be Redeemed?" 205.

98. Ibid., 211.

99. Ibid., 208.

100. Rolston discusses cruciform nature in a number of his works, including *Science and Religion*, *Genes, Genesis, and God*, and various other essays dealing with suffering and disvalues.

101. Rolston, *Science and Religion*, 133.

102. Rolston, *Genes, Genesis, and God*, 306.

103. Ibid., 307.

104. Rolston, *Science and Religion*, 145.

105. Rolston, *Genes, Genesis, and God*, 305.

106. Darwin, *Autobiography of Charles Darwin*, 88–89.

107. Ibid.

108. Ibid., 90.

109. Rolston, *Genes, Genesis, and God*, 305.

110. By "present" condition, I'm excluding anthropogenic, large-scale conditions of environmental destruction. I do not mean to imply that God intends for the humans to pollute the environment or endanger species, but rather that God willed that life processes (ecological, evolutionary processes) operate as they do.

111. Attfield, "Evolution, Theodicy, and Value," 293.

112. Rolston, "Biology and Philosophy in Yellowstone," 243.

113. Rolston, "Wildlife and Wildlands," 131.

114. Ibid., 131.

115. Leopold, *A Sand County Almanac*, 110.

116. In fact, the relationship between the scientific "picture" of nature and the ethic that it supports is not quite so straightforward, since it appears that a theological assumption about nature's fallen status influences the way in which scientific data is interpreted. Thus, for instance, McFague's ecological model is shaped more by eschatological hopes than by ecological data.

117. Gustafson, *A Sense of the Divine*, 47.

118. Gustafson, *Ethics from a Theocentric Perspective*, 1:88.

119. Ibid., 1:18.

120. Ibid., 1:19.

121. Ibid., 1:97.

122. Gustafson, *A Sense of the Divine*, 68.

123. Gustafson, *Ethics from a Theocentric Perspective*, 1:85.

124. Ibid., 2:289.

125. Land ethics, too, emphasizes this point about the necessity of moral action and illustrates why "letting nature be" is not a serious alternative.

126. Gustafson, *A Sense of the Divine*, 47.

127. Gustafson, *Ethics from a Theocentric Perspective*, 1:195.

128. Ibid., 1:211.

129. Ibid.

130. Gustafson, *A Sense of the Divine*, 44.

131. Rolston, *Conserving Natural Values*, 159. It is unfortunate that Rolston (and Gustafson, for that matter) does not manage to avoid *andro*centric language in his attempt to avoid anthropocentric perspectives!

132. Gustafson, *Ethics from a Theocentric Perspective*, 1:82.

133. Ibid., 1:83.

134. McFague, "Imaging a Theology of Nature," 203.

135. Gustafon, *Ethics from a Theocentric Perspective*, 1:60.

136. Gustafson, *A Sense of the Divine*, 72.

137. Gustafson, *Ethics from a Theocentric Perspective*, 1:58.

138. Ibid., 1:61.

139. Northcott, *The Environment and Christian Ethics*, 150.

140. Gustafson, *Ethics from a Theocentric Perspective*, 1:100.

141. Birch and Cobb, *The Liberation of Life*, 161.

142. This difference follows from the fact that although animals have richness of experience and some form of psychological life, their mental powers are aimed primarily toward bodily existence (see chapter 3). Obligations to animals tend to involve their physical suffering, even though Birch and Cobb do not deny that (some) animals have a rich subjective life.

143. Birch and Cobb, *The Liberation of Life*, 182.

144. Gustafson, *Ethics from a Theocentric Perspective*, 1:73.

145. Ibid., 1:74.

146. Ibid., 1:84 (my emphasis).

147. Ibid., 1:23.

148. McFague, *Super, Natural Christians*, 174.

149. Northcott, *The Environment and Christian Ethics*, 137.

150. Gustafson, *Ethics from a Theocentric Perspective*, 1:83.

151. Moltmann, *God in Creation*, 196.

152. Ibid., 197.

153. Ibid., 138, 13.

154. Ibid., 197.

155. John Haught's recent work, *God After Darwin*, also takes an "eschatological-evolutionary ecological" approach. Although Haught's main objective in this work is to formulate a scientifically accurate evolutionary theology (as well as an evolutionary *ecolog-*

ical theology), he arrives at an understanding of Darwinian nature that is, in my view, disappointing in its hope for nature's "redemption" and "perfection." He suggests that we think of nature as more a "promise" than a "gift" (the latter being too instrumental and unrealistic an interpretation of nature's resources for us). Viewing nature as promise leads us to understand it as having "intrinsic, but by no means ultimate, value." Ibid., 157. Furthermore, understanding nature as a promise allows us to accept its unfinished nature and its limitations "in light of the conviction that full perfection lies only in the eschatological future." Ibid., 158. This theological vision also makes it possible for us "to face up" to the suffering and ambiguity that evolution entails. Ibid., 158.

However, I fail to see how Haught's ecotheology constitutes a significant improvement over the way that Christians have traditionally viewed nature—namely, as a corrupted, fallen realm awaiting divine restoration. Nature's imperfection and ambiguity is not really "faced up to" in this theology—it is simply acknowledged until such time as it is finally removed. In not having "ultimate" value, nature is also not the ultimate reality. We merely tolerate nature's shortcomings "including instances where it seems indifferent to us" until such time as the promise is fulfilled, ambiguity removed, and tragedy redeemed. Here Haught concurs with Moltmann in his view of evolution as a transitory stage prior to nature's ultimate perfection. He also begins to sound like McFague when he argues that our "model of ecological concern" ought to be the life of Jesus who cared above all for the "relationless: the sinners, the religiously despised, the sick—and the dead." Ibid., 163. A Darwinian and Christian ecological theology "extends Jesus' inclusive compassion for the unincluded toward all of nature." Ibid., 164. This theology is more Christian than evolutionary and does not offer any profound insights beyond those already made by contemporary Christian ecotheologians, despite Haught's criticism of the pre-Darwinian orientation of current theology.

156. Gustafson, *Ethics from a Theocentric Perspective*, 1:112.

157. Gustafson, *A Sense of the Divine*, 44.

158. Ibid.

159. Gustafson, *Ethics from a Theocentric Perspective*, 1:25

160. Ibid.

161. Ibid., 1:210.

6. A COMPREHENSIVE NATURALIZED ETHIC

1. Haught, *God After Darwin*, 2.

2. Ecotheologians are enamored of developments in physics that appear to undermine dualistic, mechanistic causality. In this sense the "ecological model" seems at least partially derived from physics rather than biology. Many ecotheologians point to arguments in the physical sciences that seem to promote an understanding of all things (even nonliving entities) as interdependent and interconnected. In fact the inspiration for the ecological model comes less from the biological sciences than from physics and has been imported from physics because it seems to express an attitude toward nature that ecotheologians believe is a preferable one (to, say, the mechanistic model, which was also derived from physics.) Birch and Cobb, for instance, appropriate from physics the idea of "event thinking" (in lieu of "substance thinking") to support the idea of ecological interdependence: "Field theory, relativity physics and quantum mechanics all point in the direction of event thinking instead of substance thinking. We believe the same is true of biology and that this is witnessed to in much that is now being said and done." Birch and Cobb, *The Liberation of*

Life, 87. While they eschew the sort of reductionism that would explain animal behavior with exclusive reference to events at the molecular level, they believe that the complex interconnections found in electromagnetic "events" have implications for "animal behavior in the wild." Ibid., 89. Field theory demonstrates that "events cannot exist apart from the field" just as animals are inseparable from their environments. In short, the concept of the interdependence of all events—the inextricable links between all events and their environments—is borrowed from physics in order to support an ecological ethic. Here again the ecological model seems less derived from a study of ecology and evolution than it is extended to these realms from some other source. Once interdependence is established as a scientific fact, it is adopted as an ethic that counters dualistic, oppressive, and objectifying attitudes toward the environment—an ethic, in other words, of liberation.

In their enthusiasm to promote the idea of nature's interdependence, ecotheologians seem to forget where the insight of "interdependence" originally came from. Birch and Cobb offer a brief history of the Newtonian mechanistic model, tracing its dangerous spread to biological sciences, and showing how it has ultimately been called into question by subsequent developments in physics. With this new and better model from physics in mind, they then make a case for adopting the new model in biology: "It is now time to seek the explanation of behaviour at one level in terms of behaviour at other levels." Ibid., 86. However, they then proceed to discuss the ecological model as though it emerged in biology and can now be extended to nonliving things, arguing that "the model proposed for the study of living things is actually appropriate to the study of non-living things as well." Ibid., 93. McFague too borrows the ecological model from physics, but then seems to forget that this move has occurred: "In the early years of the twentieth century," she writes, "there was a movement toward a model more aptly described as organic, even for the constituents with which physics deals, for there occurred a profound realization of the deep relations between space, time, and matter, which relativized them all." McFague, *Models of God*, 11. This new, nondualistic perspective, which she terms the ecological model, replaced the "Newtonian" picture. Given its source, it seems more appropriate to dub this model the "field model," the "electromagnetic model," or the "relativity model," rather than the ecological model.

3. Although the concept of a food chain originated with Cambridge zoologist Charles Elton in the 1920s, the popular acceptance of the idea, I suspect, owes much to Rachel Carson and the disturbing picture she presents in *Silent Spring* of pesticides spreading through an entire community by means of these links between organisms. Carson describes the most "sinister feature of DDT"—its ability to be "passed on from one organism to another through all the links of the food chains": "Fields of alfalfa are dusted with DDT; meal is later prepared from the alfalfa and fed to hens; the hens lay eggs which contain DDT. Or the hay . . . may be fed to cows. The DDT will turn up in the milk in the amount of about 3 parts per million, but in butter made from this milk the concentration may run to 65 parts per million. . . . The poison may also be passed on from mother to offspring. . . . This means that the breast-fed human infant is receiving small but regular additions to the load of toxic chemicals building up in his body." Carson, *Silent Spring*, 23.

4. Darwin, *The Origin of Species*, 83.

5. Interdependence in this sense is harder to derive from the physics, field theory, model. Genealogical interdependence is clearly more appropriate in biology than physics and may more legitimately be referred to as a feature of the ecological model, although in

this instance the phrase "ecological model" is still somewhat questionable, given that the sort of interdependence expressed is more precisely a function of evolutionary processes.

6. McFague, *Super, Natural Christians*, 21 (my emphasis). See Lloyd and Keller, *Keywords in Evolutionary Biology*, for a discussion of the many different, even contradictory, ways in which terms such as *community, mutualism*, and *competition* are used in discussions of evolutionary biology.

7. Ruether, *Gaia and God*, 47.

8. McFague, *Super, Natural Christians*, 108.

9. Ruether, *Gaia and God*, 4.

10. Darwin, *The Origin of Species*, 84.

11. Ibid., 74.

12. Ibid.

13. We can see why Moltmann's concept of created niches and food supplies for each "kind" of organism is naive, since the struggle for existence is the most severe among members of the same species. The fact that species will have a common food supply and geographic range (as his concept of ecology emphasizes) increases, rather than decreases, struggle within "kinds."

14. Darwin, *The Origin of Species*, 75.

15. Ibid., 79.

16. Gustafson, *Ethics from a Theocentric Perspective*, 1:106.

17. Ibid., 2:284.

18. Gustafson, *A Sense of the Divine*, 48.

19. Ibid., 55.

20. Ibid., 104.

21. Gustafson, *Ethics from a Theocentric Perspective*, 2:239–43.

22. Gustafson, *A Sense of the Divine*, 106.

23. Ibid., 67.

24. Ibid., 72.

25. Ibid., 12.

26. Gustafson's comment appears in the panel discussion in Beckley and Swezey, *James M. Gustafson's Theocentric Ethics*, 227.

27. Gustafson, *A Sense of the Divine*, 49.

28. Ibid., 135.

29. Ibid., 55.

30. Ibid., 62.

31. Gustafson refers to respect as a response to the fact of dependence, as when he argues that dependence calls for a disposition of respect. But he also refers to dependence as a fact in itself. He does not draw a clear line between the fact and the response at all times (since his theology makes dispositions, affectivities more central than rational responses, facts are not easily separated from the feelings they generate) but he *does* draw a line between the response and the specific moral action that is taken (i.e., between response and responsibility).

32. Garrison, "Moral Obligations to Non-Human Creation," 204. Garrison locates several parallels between Gustafson's theocentric ethics and Rolston's nonanthropocentric environmental ethics, both in terms of their methods and their substantive claims. Both Rolston and Gustafson, for example, understand the "is/ought" relationship to be more

complex and ambiguous than many ethicists and theologians realize. Both look to nature for "indicators" of appropriate human action, rather than locating values in nature in a simple, deductive manner. Moreover, Garrison correctly identifies similarities between Rolston's conception of systemic value (which places emphasis on natural processes and not merely products) and Gustafson's account of multidimensional, theocentric values that are "affected by relationships." Ibid., 219.

33. Callicott, "Animal Liberation," 58.

34. Gustafson, *A Sense of the Divine*, 46–47.

35. Scoville, "Leopold's Land Ethic," 58.

36. Scoville maintains that Callicott has overstated the ecocentric orientation of Leopold's ethic, creating the impression that land ethics is "unconcerned about human welfare." Ibid., 62.

37. Northcott, *The Environment and Christian Ethics*, 109.

38. Scoville, "Leopold's Land Ethic," 70.

39. Callicott, "Animal Liberation," 54.

40. Ibid., 54.

41. Gustafson, *Ethics from a Theocentric Perspective*, 1:283.

42. Rolston, *Conserving Natural Value*, 114.

43. Gustafson, *Ethics from a Theocentric Perspective*, 1:283.

44. Gustafson, *A Sense of the Divine*, 48.

45. Gustafson, *Ethics from a Theocentric Perspective*, 1:244.

46. Leopold, *A Sand County Almanac*, 205.

47. McFague, *Super, Natural Christians*, 106.

48. Ibid., 110.

49. Ibid., 15.

50. Darwin, *The Descent of Man*, 1:39.

51. Ibid., 1:41–42.

52. Ibid., 1:40.

53. As noted in chapter 1, however, Darwin was critical of experimentation on animals without anesthesia.

54. Wilson, "Biophilia and the Conservation Ethic," 33.

55. See Heerwagen and Orians, "Humans, Habitats, and Aesthetics," 138.

56. Kellert, "The Biological Basis," 50.

57. Wilson, "Biophilia and the Conservation Ethic," 39.

58. A *conservation* ethic is Wilson's term for the type of ethic toward nature we need. He means by this a broad ethic that would protect biodiversity as a whole.

59. Soule, "Biophilia," 442.

60. Wilson, "Biophilia and the Conservation Ethic," 31.

61. Ibid., 38.

62. Soule, "Biophilia," 446.

63. Wilson, "Biophilia," 39. Not all aesthetic responses can currently be explained by the biophilia hypothesis, Wilson would concede, but he is convinced that their functional explanations can be found in our evolutionary heritage.

64. Ibid., 37 (my emphasis).

65. Midgley, *Animals and Why They Matter*, 102.

66. Ibid., 104.

67. Terborgh, "In the Company of Humans," 61.

68. Ibid., 62.

69. Ibid.

70. Ecotheologians' quest for a thoroughly nonobjectifying, noninstrumentalizing relationship between humans and other animals seems misguided in light of evolutionary considerations of how our relationships with them were originally formed. Terborgh's argument also suggests that an ethic such as Regan's, that eschews any and all instrumentalization of beings who are subjects, may incorrectly assume that instrumentalizing other beings is an unnatural, human aberration.

71. Terborgh, "In the Company of Humans," 62.

72. Midgley, *Beast and Man*, xxxix.

73. Ibid., 53.

74. Ibid., 54.

75. Midgley, *Animals and Why They Matter*, 111.

76. See Budiansky, *The Covenant of the Wild*, for a discussion of the evolutionary basis for domestication of animals.

77. Midgley, *Animals and Why They Matter*, 117.

78. Ibid., 119.

79. Wilson, *Biophilia*, 21.

80. Ibid., 22.

81. Ibid., 2.

82. Terborgh, "In the Company of Humans," 62.

83. Ibid., 62.

84. Wilson, *Biophilia*, 140.

85. Ibid., 131.

86. Rolston, "Biophilia, Selfish Genes, Shared Values," 381.

87. Ibid., 411.

88. Rolston, "Wildlife and Wildlands," 123.

89. Rolston, "Biophilia, Selfish Genes, Shared Values," 409.

90. But does the idea that humans are participants and not controllers of nature contradict the image proposed by Rolston of humans as "overseers" in the natural world? Rolston's argument in particular (more so than, say, Midgley's) raises certain questions because of his clearer separation between nature and culture. In proposing that humans are "cognitive," "ethical," and "cultural" beings who now live more or less outside of nature selection, is he not simply reasserting an "image-of-God" model of humans, or perhaps even a dualistic, Cartesian argument that humans, possessing rational minds, are given the task of ordering physical, nonrational nature?

Admittedly, Rolston sometimes veers too close to this sort of interpretation—and has been criticized even by other land ethicists for drawing too sharp a line between humans and nature. At one point he suggests that our ability to see the larger picture of nature confers to us a status along the lines of "prophets, priests, and kings. . . . Humans should speak for God in natural history." Rolston, "Wildlife and Wildlands," 124. I confess to finding these elements of Rolston's argument troubling, infrequent though they appear to be. It is often difficult to know, given the context in which such statements appear, whether he is simply rehearsing the "traditional" Christian view in order to revise it in light of his own ethics or whether he believes that such statements can be reconciled with a land ethics perspective. He is quick to add, for instance, that there is "growing conviction that theology has been too anthropocentric." Ibid., 124. The bulk of Rolston's argument sounds considerably more theocentric and, in many places, remarkably similar to Gustafson's.

91. Rolston, "Biophilia, Selfish Genes, Shared Values," 409.

92. Ibid., 410.

93. Clearly, there are differences as well between these interpretations of a basic response. Gustafson and land ethicists, for instance, respect processes themselves as much if not more so than the products. They also do not necessarily think of this response as one of love toward nature (or in Gustafson's case, toward the deity). Ecotheologians understand the response in keeping with Christian love and the paradigm of liberation, and some value individual products (animals, etc.) rather than the processes themselves. Nevertheless, all express a general inclination to respond morally to nature.

94. Leopold, *A Sand County Almanac*, 223.

95. Even Gustafson, who does not often use language of love for nature, insists that the response is an affective one—hence, his criticism of purely rational appraisals of nature's interdependence. He is interested in religious "feelings" about the natural ordering. Both he and Midgley, moreover, are critical of contractarian and Kantian (rational) accounts of moral obligations and duties.

96. Midgley, *Beast and Man*, 58.

97. Ibid., 274.

98. Ibid., 270.

99. Ibid., 168.

100. Butler, *Sermons*.

101. Midgley, *Beast and Man*, 282.

102. Ibid., 282.

103. Gustafson, *Ethics from a Theocentric Perspective*, 1:299. Stephen Pope makes a similar argument for a "comprehensive" biological and theological perspective on nature. "Appropriations of behavioral biology must proceed in a nonreductionist way, in full recognition of the fact that natural selection is but one of several factors." Pope, *The Evolution of Altruism*, 121. Like Gustafson, Pope cites Midgley's work as an example of such an approach. He proposes a greater open-endedness to human behavior than selfish genes or biophilia dictate, arguing that we are "cultural animals with 'open programs' . . . able to extend moral concern (at least in the sense of nonmaleficence) to all human beings, even though the human biological substrate may not strongly incline people to act in this direction." Ibid., 140. Darwin himself presented a similar case for moral extension, as Pope correctly points out, predicting that human concern, which began with a set of social instincts we share with other animals, would ultimately expand outward to include all other humans as well as animals. Intellectual powers would aid in this extension.

104. Webb, *On God and Dogs*, 6. Webb argues that the human-pet relationship is distinguished by a degree of unconditional love that even our relationships with other humans rarely attain—an unconditional love that is, incidentally, symbolic of the human-divine relationship. "My primary thesis," he writes, "is that the very excess of the human-dog relationship is the key to a way of relating to animals that is often overlooked or dismissed in serious, scholarly treatments of animals." Unfortunately, however, Webb does not limit "animals" here to *nonwild* animals.

105. Midgley, *Animals and Why They Matter*, 101.

106. McFague, *Super, Natural Christians*, 113.

107. National Research Council, *Science and the Endangered Species Act*, 37.

108. Rolston, *Environmental Ethics*, 66.

109. Midgley, *Beast and Man*, 351.

110. In *God After Darwin*, Haught takes issue with the language of "gifts," preferring instead the idea of nature as "promise," insisting that the gift concept sounds too instrumental. Within a theocentric context, however, the instrumental implications are greatly mitigated: the gift is not anthropocentrically valued and the proper response is one of gratitude rather than exploitation. Indeed, "promise" in the context of Haught's work has eschatological overtones which I find far more problematic.

111. Rolston, "Wildlife and Wildlands," 135.

112. Ibid., 136.

113. National Research Council, *Science and the Endangered Species Act*, 113–114.

114. Ibid., 120. In such cases, ecotheologians are correct in arguing that humans have *created* "conflict" in nature. But the remedy they suggest—often based upon disproportionately valuing and protecting individual "subjects" or highly "rich" organisms, would only exacerbate these conflicts. Moreover, even once such conflicts are remedied, if they ever are, conflict itself (between predator and prey, for instance) remains a constant in nature.

115. Rolston, *Conserving Natural Value*, 114.

116. Attfield, "Rehabilitating Nature and Making Nature Habitable," 49.

117. Ibid., 50.

118. See Berg and Dasmann, "Reinhabiting California," 217–220. A more recent anthology is Jackson and Vitek, *Rooted in the Land*.

119. Leopold, *A Sand County Almanac*, 204.

120. See Quinn, *How Wal-Mart Is Destroying America*.

121. McCloskey, "Ecology and Community."

122. See Cafaro, "Thoreau, Leopold, Carson."

123. Callicott, "Animal Liberation and Environmental Ethics," 257.

124. Rolston, "Wildlife and Wildlands," 132.

125. This is not to say that we should not aim at reduction or even elimination of animals kept for food. This relationship, in all but its most humane forms (small, free-range farms) is a violation of the relationship we have entered into with domestic animals. There are very good environmental reasons for eliminating meat from our diet as well, owing to the amount of land and water that is currently being wasted to support livestock, not to mention the incredible amounts of animal waste discharged into our environment every day. But I see no similar moral objection to keeping animals as pets, so long as they are treated humanely and their impact on wild nature is minimal.

126. Andrew Linzey has proposed the language of "parenthood" applied to "human relations with animals," particularly pets: "Whenever we find ourselves in a position of power over those who are relatively powerless our moral obligation of generosity increases in proportion." Linzey, *Animal Theology*, 37–38. Like McFague, he gives moral priority to the weak, drawing on the example of Jesus. Linzey adopts a rights position, however, that does not draw clear enough lines between wild and domestic animals.

127. Midgley, *Beast and Man*, 165.

WORKS CITED

Alberta Agriculture, Food, and Rural Development. "Ginseng Industry." November 1996. http://www.agric.gov.ab.ca/agdex/100/8883002.html.

Attfield, Robin. "Evolution, Theodicy, and Value." *Heythrop Journal* 41 (2000).

——— "Rehabilitating Nature and Making Nature Habitable." In Robin Attfield, ed., *Philosophy and the Natural Environment*. Cambridge: Cambridge University Press, 1994.

Augustine, *City of God*. Ed. Henry Bettenson, with and introduction by David Knowles. 4th ed. Hammondsworth: Penguin, 1980.

Barbour, Ian. *Religion and Science: Historical and Contemporary Issues*. Rev. ed. San Francisco: HarperCollins, 1997.

Beckley, Harlan R. and Charles M. Swezey, eds., *James M. Gustafson's Theocentric Ethics: Interpretations and Assessments*. Macon: Mercer University Press, 1988.

Bentham, Jeremy. *An Introduction to the Principles of Morals and Legislation*. Oxford: Clarendon, 1789.

Berg, Peter and Raymond Dasmann. "Reinhabiting California." In Peter Berg, ed., *Reinhabiting a Separate Country: A Bioregional Anthology of Northern California*. San Francisco: Planet Drum, 1978.

Birch, Charles. "Christian Obligation for the Liberation of Nature." In Charles Birch and William Eakin, eds., *Liberating Life: Contemporary Approaches to Ecological Theology*. Maryknoll: Orbis, 1990.

Birch, Charles and John Cobb. *The Liberation of Life: From the Cell to the Community*. Cambridge: Cambridge University Press, 1981.

Botkin, Daniel. *Discordant Harmonies: The New Ecology of the Twenty-first Century*. Oxford: Oxford University Press, 1990.

Bowler, Peter. *Evolution: The History of an Idea*. Berkeley: University of California Press, 1984.

——— *The Non-Darwinian Revolution: Reinterpreting a Historical Myth*. Baltimore: Johns Hopkins, 1988.

Budiansky, Stephen. *The Covenant of the Wild: Why Animals Chose Domestication*. New Haven: Yale University Press, 1999.

Butler, Joseph. *Sermons*. Ed. W. R. Matthews. London: Bell, 1969 [1726].

Cafaro, Philip. "Thoreau, Leopold, and Carson: Toward an Environmental Virtue Ethics." *Environmental Ethics* 22 (Spring 2001): 3–17.

Callicott, J. Baird. "Animal Liberation: A Triangular Affair." In Robert Elliot, ed., *Environmental Ethics*, 29–59. Oxford: Oxford University Press, 1995.

——— "Animal Liberation and Environmental Ethics: Back Together Again." In Eugene Hargrove, ed., *The Animal Rights/Environmental Ethics Debate*, 249–61. Albany: SUNY Press, 1992.

——— "Can a Theory of Moral Sentiments Support a Genuinely Normative Environmental Ethic?" *Inquiry* 35 (1992): 183–98.

——— "The Conceptual Foundations of the Land Ethic." In Joseph DesJardins, ed., *Environmental Ethics: Concepts, Policy, and Theory*, 227–37. Mountain View: Mayfield, 1999.

——— *In Defense of the Land Ethic: Essays in Environmental Philosophy*. Albany: State University of New York Press, 1989.

——— "Do Deconstructive Ecology and Sociobiology Undermine Leopold's Land Ethic?" *Environmental Ethics* 18 (1996): 359–72.

——— "Rolston on Intrinsic Value: A Deconstruction." *Environmental Ethics* 14 (1992): 129–43.

——— "The Wilderness Idea Revisited." In J. Baird Callicott and Michael Nelson, eds., *The Great New Wilderness Debate*. Athens: University of Georgia Press, 1998.

Carson, Rachel. *Silent Spring*. New York: Houghton Mifflin. 1962.

Chase, Alston. *Playing God in Yellowstone*. New York: Harcourt Brace Jovanovich, 1987.

Cobb, John, and David Griffin. *Process Theology*. Philadelphia: Westminster, 1976.

Cobbe, Francis Power. *Life of Francis Power Cobbe*. 2 vols. Boston: Houghton Mifflin, 1894.

Cronon, William. "The Trouble with Wilderness; or, Getting Back to the Wrong Nature." In William Cronon, ed., *Uncommon Ground: Rethinking the Human Place in Nature*, 69–90. New York: Norton, 1996.

Darwin, Charles. *Autobiography of Charles Darwin*. Ed. Nora Barlow. New York: Norton, 1958.

——— *The Descent of Man and Selection in Relation to Sex*. Princeton: Princeton University Press, 1981.

——— *The Life and Letters of Charles Darwin*. Ed. Francis Darwin. 3 vols. London: John Murray, 1888.

——— *The Origin of Species*. London: Penguin, 1987.

Dawkins, Richard. *The Selfish Gene*. Oxford: Oxford University Press, 1976.

Dennett, Daniel. *Darwin's Dangerous Idea*. New York: Touchstone, 1995.

Descartes, René. "Discourse on Method." *Philosophical Works of Descartes*. Ed. E. S. Haldane and G. R. T. Ross. 2 vols. London: Cambridge University Press, 1931.

Fitzsimmons, Allen. "Ecological Confusion Among the Clergy. *Journal of Markets and Morality* 3.2 (Fall 2000): 204–23.

Fleming, Donald. "Charles Darwin, the Anaesthetic Man." *Victorian Studies* 4 (March 1961): 219–36.

Fowler, Robert Booth. *The Greening of Protestant Thought*. Chapel Hill: University of North Carolina Press, 1995.

Garrison, Glenn G. "Moral Obligations to Non-Human Creation: A Theocentric Ethic." Ph.D. diss., Southern Baptist Theological Seminary, 1994.

Golley, Frank Benjamin. *A History of the Ecosystem Concept in Ecology: More than the Sum of the Parts*. New Haven: Yale University Press, 1993.

Goodwin, Brian. *How the Leopard Changed Its Spots: The Evolution of Complexity*. London: Phoenix, 1997.

Gould, Stephen Jay. "Darwinism Defined." *Discover* (January 1987): 64–70.

——— *Ever Since Darwin*. New York: Norton, 1979.

——— "Fulfilling the Spandrels of World and Mind." In Jack Selzer, ed., *Understanding Scientific Prose*, 320–36. Madison: University of Wisconsin Press, 1993.

——— *The Mismeasure of Man*. New York: Norton, 1981.

——— *The Panda's Thumb*. New York: Norton, 1982.

——— *Rocks of Ages: Science and Religion in the Fullness of Life*. New York: Ballantine, 1999.

——— *The Structure of Evolutionary Theory*. Cambridge: Belknap, 2002.

——— *Wonderful Life: The Burgess Shale and the Nature of History*. New York: Penguin, 1991.

Gould, Stephen Jay and Richard Lewontin. "The Spandrels of San Marco and the Panglossian Paradigm: A Critique of the Adaptationist Programme." *Proceedings of the Royal Society of London* (B) 205 (1978): 581–98.

Gregersen, Niels Henrik. "The Cross of Christ in an Evolutionary World." *Dialog: A Journal of Theology* 40.3 (2001): 192–207.

Gustafson, James. *Ethics from a Theocentric Perspective*. 2 vols. Oxford: Basil Blackwell, 1981.

——— *A Sense of the Divine: The Natural Environment from a Theocentric Perspective*. Cleveland: Pilgrim, 1994.

Hagen, Joel. *An Entangled Bank: The Origins of Ecosystem Ecology*. New Brunswick: Rutgers University Press, 1992.

Haught, John. *God After Darwin*. Boulder: Westview, 2000.

Heerwagen, Judith and Gordon H. Orians. "Humans, Habitats, and Aesthetics." In Stephen Kellert and E. O. Wilson, eds., *The Biophilia Hypothesis*, 138–72. Washington, D.C.: Island, 1993.

Hull, David L. "Activism, Scientists, and Sociobiology." *Nature* 407 (2000): 673–674.

Jackson, Wes and William Vitek, eds. *Rooted in the Land: Essays in Community and Place*. New Haven: Yale University Press, 1996.

Kellert, Stephen, "The Biological Basis for Human Values of Nature." In Stephen Kellert and E. O. Wilson, eds., *The Biophilia Hypothesis*, 42–69. Washington, D.C.: Island, 1993.

Kirby, Alex. "Plea to Cherish Mini-Beasts." *BBC News Online*. June 5, 2000.

Le Blanc, Jill. "A Mystical Response to Disvalue in Nature." *Philosophy Today* 45.3 (Fall 2001): 254–65.

Leopold, Aldo. *For the Health of the Land: Previously Unpublished Essays and Other Writings*. Ed. J. Baird Callicott and Eric Freyfogle. Washington, D.C.: Island, 1999.

——— "The Land Health Concept and Conservation." In J. Baird Callicott and Eric Freyfogle, eds., *For the Health of the Land: Previously Unpublished Essays and Other Writings*. Washington, D.C.: Island, 1999.

——— "Planning for Wildlife." In J. Baird Callicott and Eric Freyfogle, eds., *For the Health of the Land: Previously Unpublished Essays and Other Writings*. Washington, D.C.: Island, 1999.

——— *A Sand County Almanac*. Oxford: Oxford University Press, 1949.

——— "Some Fundamentals of Conservation in the Southwest." *Environmental Ethics* 1 (1979 [1923]): 131–41.

——— "What Is a Weed?" In J. Baird Callicott and Eric Freyfogle, eds., *For the Health of the Land: Previously Unpublished Essays and Other Writings*, 207–12. Washington, D.C.: Island, 1999.

Levins, Richard and Richard Lewontin. *The Dialectical Biologist.* Cambridge: Harvard University Press, 1985.

Linnaeus, Carolus. "The Oeconomy of Nature." In Owen Goldin and Patricia Kilroe, eds., *Human Life and the Natural World.* Trans. Benjamin Stillingfleet. Peterborough, Ontario: Broadview, 1997 [1749].

Linzey, Andrew. *Animal Theology.* Urbana: University of Illinois Press, 1994.

Lloyd, Elisabeth and Evelyn Fox Keller, eds. *Keywords in Evolutionary Biology.* Harvard: Harvard University Press, 1992.

Lorbiecki, Marybeth. *A Fierce Green Fire.* Helena: Falcon, 1996.

Lovelock, James. *The Ages of Gaia.* New York: Bantam, 1990.

Lovelock, James and Sidney Epton. "In Quest for Gaia." In Louis Pojman, ed., *Environmental Ethics*, 142–46. Boston: Jones and Bartlett, 1994.

McCloskey, David. "Ecology and Community: The Bioregional Vision." www.tnews.com/text/mccloskey.2html.

McFague, Sallie. "Imaging a Theology of Nature." In Charles Birch and William Eakin, eds., *Liberating Life: Contemporary Approaches to Ecological Theology*, 201–27. Maryknoll: Orbis, 1990.

——— *Life Abundant: Rethinking Theology and Economy for a Planet in Peril.* Minneapolis: Fortress, 2001.

——— *Metaphorical Theology: Models of God in Religious Language.* Philadelphia: Fortress, 1982.

——— *Models of God: Theology for an Ecological, Nuclear Age.* Philadelphia: Fortress, 1987.

——— *Super, Natural Christians.* Minneapolis: Fortress, 1997.

McKibben, Bill. *The End of Nature.* Garden City, N.J.: Anchor, 1990.

Margulis, Lynn and Dorion Sagan. "God, Gaia, and Biophilia." In Stephen Kellert and E. O. Wilson, eds., *The Biophilia Hypothesis*, 345–64. Washington, D.C.: Island, 1993.

Marshall, Peter. *Nature's Web.* London: Sharpe, 1992. http://www.tnews.com/text/mccloskey2.html.

Merchant, Carolyn. *The Death of Nature: Women, Ecology, and the Scientific Revolution.* San Francisco: Harper Collins, 1983.

Midgley, Mary. *Animals and Why They Matter.* Athens: University of Georgia Press, 1983.

——— *Beast and Man: The Roots of Human Nature.* New York: Routledge, 1979.

——— "The Paradox of Humanism." In Harlan R. Beckley and Charles M. Swezey, eds., *James M. Gustafson's Theocentric Ethics: Interpretations and Assessments.* Macon: Mercer University Press, 1988.

——— *Science as Salvation: A Modern Myth and Its Meaning.* New York: Routledge, 1992.

——— "The Significance of Species." In Eugene Hargrove, ed., *The Animal Rights/Environmental Ethics Debate.* Albany: SUNY Press, 1992.

Moltmann, Jurgen. *God in Creation.* Minneapolis: Fortress, 1993.

Nash, Roderick. *The Rights of Nature: A History of Environmental Ethics.* Madison: University of Wisconsin Press, 1989.

National Research Council. *Science and the Endangered Species Act*. Washington, D.C.: National Academy Press, 1995.
Nixon, Will. "The Species Only a Mother Could Love." *Amicus Journal* (Summer 1999).
Northcott, Michael. *The Environment and Christian Ethics*. Cambridge: Cambridge University Press, 1996.
Numbers, Ronald. *Darwinism Comes to America*. Cambridge: Harvard University Press, 1998.
Odum, Eugene. *Fundamentals of Ecology*. Philadelphia: Saunders, 1953.
Pope, Stephen. *The Evolution of Altruism and the Ordering of Love*. Washington, D.C.: Georgetown University Press, 1994.
Quinn, Bill. *How Wal-Mart Is Destroying America: And What You Can Do About It*. Berkeley: Ten Speed, 1998.
Rachels, James. *Created from Animals: The Moral Implications of Darwinism*. Oxford: Oxford University Press, 1990.
Rasmussen, Larry. *Earth Community: Earth Ethics*. Orbis: Maryknoll. 1996.
Regan, Tom. "Animal Rights, Human Wrongs." In Michael E. Zimmerman, ed., *Environmental Philosophy: Animal Rights to Radical Ecology*, 33–48. Englewood Cliffs: Prentice-Hall, 1993.
——— *The Case for Animal Rights*. Berkeley: University of California Press, 1983.
———"Christianity and Animal Rights." In Charles Birch and William Eakin, eds., *Liberating Life: Contemporary Approaches to Ecological Theology*, 73–87. Maryknoll: Orbis, 1990.
——— "The Radical Egalitarian Case for Animal Rights." In Louis Pojman, ed., *Environmental Ethics*. Boston: Jones and Bartlett, 1994.
Rolston, Holmes, III. "Biology and Philosophy in Yellowstone." *Biology and Philosophy* 5 (1990): 241–58.
——— "Biophilia, Selfish Genes, Shared Values." In Stephen Kellert and E. O. Wilson, eds., *The Biophilia Hypothesis*, 381–414. Washington, D.C.: Island, 1993.
——— "Challenges in Environmental Ethics." In Michael E. Zimmerman, ed., *Environmental Philosophy: Animal Rights to Radical Ecology*, 135–57. Englewood Cliffs: Prentice Hall, 1994.
——— *Conserving Natural Value*. New York: Columbia University Press, 1994.
——— "Disvalues in Nature." *Monist* 75 (1992): 250–78.
——— "Does Nature Need to be Redeemed?" *Zygon* 29.2 (1994): 205–29.
——— "Duties to Endangered Species." In Robert Elliot, ed., *Environmental Ethics*, 60–7. Oxford: Oxford University Press, 1990.
——— *Environmental Ethics: Duties to and Values in the Natural World*. Philadelphia: Temple University Press, 1988.
——— "Environmental Ethics in Antarctica." *Environmental Ethics* 24 (2002): 115–34.
——— *Genes, Genesis, and God*. Cambridge: Cambridge University Press, 1999.
——— "God and Endangered Species." In Laurence Hamilton, ed., *Ethics, Religion, and Biodiversity*, 40–64. Isle of Harris: White Horse, 1994.
——— "Philosophy and Biology in Yellowstone," *Biology and Philosophy* 5 (1990): 241–58.
——— *Science and Religion: A Critical Survey*. Philadelphia: Temple University Press, 1987.
——— "The Wilderness Idea Reaffirmed." In J. Baird Callicott and Michael Nelson,

eds., *The Great Wilderness Debate*, 367–86. Athens: University of Georgia Press, 1998.

——— "Wildlife and Wildlands." In Dieter Hessel, ed., *After Nature's Revolt*, 122–43. Minneapolis: Fortress, 1992.

Ruether, Rosemary Radford. *Gaia and God: An Ecofeminist Theology of Earth Healing*. San Francisco: HarperCollins, 1994.

Ruse, Michael. *Can a Darwinian Be a Christian? The Relationship Between Science and Religion*. Cambridge: Cambridge University Press, 2001.

Ryder, Richard. *Victims of Science*. London: Poynter-Davis, 1975.

Sagoff, Mark, "Animal Liberation and Environmental Ethics: Bad Marriage, Quick Divorce." In Michael E. Zimmerman, ed., *Environmental Philosophy: Animal Rights to Radical Ecology*, 84–94. Englewood Cliffs, N.J.: Prentice Hall, 1993.

Scoville, Judith. "Leopold's Land Ethic and Ecotheology." *Ecotheology* 8 (2000): 58–70.

Segerstrale, Ullica. *Defenders of the Truth: The Battle for Science in the Sociobiology Debate and Beyond*. Oxford: Oxford University Press, 2000.

Singer, Peter. *Animal Liberation*. Rev. ed. New York: Avon, 1990.

——— "Not for Humans Only." In K. E. Goodpastor and K. M. Sayre, eds., *Ethics and Problems of the Twenty-first* Century, 191–206. Notre Dame: University of Notre Dame Press, 1979.

——— *Practical Ethics*. 2d ed. Cambridge: Cambridge University Press, 1993.

Soule, Michael. "Biophilia: Unanswered Questions." In Stephen Kellert and E. O. Wilson, eds., *The Biophilia Hypothesis*, 441–455. Washington, D.C.: Island, 1993.

Specter, Michael. "The Dangerous Philosopher." *New Yorker*, September 6, 1999.

Sterelny, Kim. *Dawkins and Gould: Survival of the Fittest*. Cambridge: Icon, 2001.

Taylor, Paul. *Respect for Nature*. Princeton: Princeton University Press, 1996.

Terborgh, John. "In the Company of Humans." *Natural History* 109.4 (2000): 54–63.

Thornhill, Randy and Craig Palmer. *A Natural History of Rape: Biological Bases of Sexual Coercion*. Cambridge: MIT Press, 2000.

Toulmin, Stephen. *The Discovery of Time*. New York: Harper and Row, 1965.

Turner, James. *Reckoning with the Beast: Animals, Pain, and Humanity in the Victorian Mind*. Baltimore: Johns Hopkins Press, 1980.

Webb, Stephen. *On God and Dogs: A Christian Theology of Compassion for Animals*. Oxford: Oxford University Press, 1998.

White, Lynn. "The Historical Roots of Our Ecological Crisis." *Science* 155 (1967): 1203–1207.

Williams, Patricia. *Doing Without Adam and Eve: Sociobiology and Original Sin*. Minneapolis: Fortress, 2001.

Wilson, Edward O. *Biophilia: The Human Bond with Other Species*. Cambridge: Harvard University Press, 1984.

——— "Biophilia and the Conservation Ethic." In Stephen Kellert and E.O. Wilson, eds., *The Biophilia Hypothesis*, 31–41. Washington, D.C.: Island, 1993.

——— *Naturalist*. Washington, D.C.: Island, 1994.

——— *Sociobiology: The New Synthesis*. Cambridge: Harvard University Press, 1975.

Worster, Donald. *Nature's Economy: A History of Ecological Ideas*. 2d ed. Cambridge: Cambridge University Press, 1994.

Wynn, Mark. "Natural Theology in an Ecological Mode." *Faith and Philosophy* 16.1 (1999): 27–42.

INDEX